Satellite Systems - Design, Modeling, Simulation and Analysis

Edited by Tien Nguyen

Published in London, United Kingdom

IntechOpen

Supporting open minds since 2005

Satellite Systems - Design, Modeling, Simulation and Analysis
http://dx.doi.org/10.5772/intechopen.73789
Edited by Tien Nguyen

Contributors
Yuri V. Kim, Igor V. Bezmenov, Tien M. Manh Nguyen, John Nguyen, Dmitry Lozhkin, Genshe Chen, Dan Shen, Carolyn Sheaff, Jingyang Lu, Mengqing Guo, Erik Blasch, Khanh Pham, Hung H. Nguyen, Peter S. Nguyen, Jack K. Kreng, Gleason Q. Qiang-Qiang Chen, Robertus Triharjanto, Xiaodong Han, Changqing Wu, Yakun Wang, Anirban Patra, Debasish Chakraborty, Arijit Saha, Kallol Bhattacharya

First published in London, United Kingdom, 2021 by IntechOpen
IntechOpen is the global imprint of INTECHOPEN LIMITED, registered in England and Wales, registration number: 11086078, 5 Princes Gate Court, London, SW7 2QJ, United Kingdom
Printed in Croatia

British Library Cataloguing-in-Publication Data
A catalogue record for this book is available from the British Library

Additional hard and PDF copies can be obtained from orders@intechopen.com

Satellite Systems - Design, Modeling, Simulation and Analysis
Edited by Tien Nguyen
p. cm.
Print ISBN 978-1-83968-373-2
Online ISBN 978-1-83968-374-9
eBook (PDF) ISBN 978-1-83968-375-6

For EU product safety concerns:
IN TECH d.o.o., Prolaz Marije Krucifikse Kozulić 3, 51000 Rijeka, Croatia, info@intechopen.com or visit our website at intechopen.com.

Meet the editor

Dr. Tien Nguyen serves as an adjunct research professor in Mathematics at California State University, Fullerton (CSUF), where he is a visiting scholar and advisor at the Center of Computational and Applied Mathematics. He works full-time as a deputy chief architect of an international program at The Aerospace Corporation, California. His previous roles at the corporation include associate director, interim direction, and principal technical staff. He was with Raytheon serving as Program-Area Chief Engineer in Advanced Concept Technology and retired as an engineering fellow in 2014. He also served as a NASA delegate to the international Consultative Committee for Space Data Systems (CCSDS), and many of his works were adopted as CCSDS standards. He received his Ph.D. in Applied Mathematics from Claremont Graduate University, California. He is an expert in advanced mathematical modeling of a complex system of systems.

Contents

Preface

There are many professional and academic books on satellite systems, satellite communications, and satellite modeling and simulation (M&S). Academic textbooks tend to focus on existing and advanced satellite systems and basic principles of satellite operations and communications. The cover topics such as satellite orbits, baseband signals, digital communication techniques, Quality of Services (QoS), uplink and downlink, link budget calculation, satellite multiple access, satellite communication payload, satellite networks, and satellite ground stations (or Earth stations). Professional satellite books tend to focus on specific satellite systems and subsystems or technologies that require the reader to have a deep technical background to understand the contents. In general, professional satellite books can be classified into three categories: advanced professional/technical books, advanced professional modeling and simulation (M&S) books, and professional handbooks focusing on the latest technological challenges.

The content of this book is a mix of topics covered in existing professional/technical books and academic textbooks. This book provides a high-level description of (1) existing and future satellite systems, (2) practical satellite system design approaches, and (3) modeling, simulation, and analysis of existing satellite systems. It is a practical reference on satellite systems and their design, modeling, simulation, and analysis for system engineers, design engineers, system analysts, and researchers in satellite engineering and advanced mathematical modeling fields.

Chapter 1 examines existing and future satellite systems, including the Legacy Analog Bent-Pipe Satellite (ABPS), Digital Bent-Pipe Satellite (DBPS), Advanced Digital Bent-Pipe Satellite using Digital Channelizer and Beamformer (AdDBPS-DCB), and future Advanced Regenerative On-Board Processing Satellite (AR-OBPS). Chapter 2 provides an end-to-end overview of existing communication subsystems and describes the function of each subsystem and its relationship to other satellite subsystems. Chapter 3 describes satellite control capabilities and the proper functioning of satellite payload(s). It discusses space control and Space Situation Awareness (SAA) capabilities, which are required to address space debris challenges. Chapter 4 presents machine learning methods with novel neural networks. Chapter 5 provides an overview of a Modular Open System Architecture (MOSA) approach for future satellite system architectures. Chapter 6 discusses two design approaches for micro-satellites currently being used by the Technical University of Berlin and the University of Surrey. Chapter 7 presents an "open" architecture along with "modular" software and hardware design for building a universal, standardized, and scalable intelligent avionics system. Chapter 8 presents a modeling, simulation, and analysis approach to assess Telemetry, Tracking, and Command (TT&C) uplink and downlink supporting launch of a Medium Earth Orbiter (MEO) satellite to final orbital insertion. Chapter 9 focuses on the development of an effective algorithm for detecting and eliminating outliers from data measurements in Global Navigation Satellite Systems (GNSS). Chapter 10 explains several levels of satellite data processing for assessing the spatiotemporal variability of surface temperature and presents analysis results for a specific region. Finally, Chapter 11 presents a novel satellite imagery data compression approach using sinusoidal amplitude grating to optimize storage space and bandwidth in satellite communications.

I would like to express my deep gratitude to all the authors and co-authors for their contributions. This book would not have been possible without their hard work as well as the support of outside reviewers and the staff at IntechOpen. I am especially grateful to Author Service Manager, Ms. Kristina Kardum Cvitan, for her help and conscientious support throughout the editing and publishing process, and my colleagues, Professors Charles Lee and Sam Behseta of California State University, Fullerton (CSUF), Mr. Leon Truong, Dr. Hung Nguyen, Mr. Andy Guillen, and Dr. Sumner Matsunaga of The Aerospace Corporation, California. Last but not least, I'm in debt to my wonderful wife, Thu-Hang Nguyen, for her constant moral support and encouragement during the process of creating this book.

Tien M. Nguyen
California State University in Fullerton (CSUF),
Adjunct Research Professor and Visiting Scholar,
Center for Computational and Applied Mathematics (CCAM)
Fullerton, California, U.S.A

Section 1

Existing and Future Satellite Systems

Chapter 1

Communication Subsystems for Satellite Design

Hung H. Nguyen and Peter S. Nguyen

Abstract

The objective of this chapter is to provide a comprehensive end-to-end overview of existing communication subsystems residing on both the satellite bus and payloads. These subsystems include command and mission data handling, telemetry and tracking, and the antenna payloads for both command, telemetry and mission data. The function of each subsystem and the relationships to the others will be described in detail. In addition, the recent application of software defined radio (SDR) to advanced satellite communication system design will be looked at with applications to satellite development, and the impacts on how SDR will affect future satellite missions are briefly discussed.

Keywords: command and data handling subsystem (C&DHS), telemetry and tracking, antenna payload, software defined radio (SDR)

1. Background and introduction

In the context of this chapter, a satellite is a spacecraft (SC) that orbits around a celestial body such as the earth. A spacecraft has several design constraints placed upon it before it can be placed in an orbit around the intended celestial body. First, satellite designs are limited in their mass and volume to fit on the launch vehicle that places them into orbit. Secondly, the mass and volume limits affect the size of the power system on the spacecraft; therefore, the amount of power available to the satellite is also limited. In addition, the space environment (thermal, radiation, atomic oxygen, space debris, micrometeoroids, etc.) imposes constraints on the design such as parts and material selection.

A spacecraft is consisted of two parts: the spacecraft bus and the payload (PL) [1, 2]. The spacecraft bus provides control of the satellite and support services to the mission payload, while the mission payload provides the mission part of the satellite including payload control, mission data processing, and mission data downlink dissemination. Examples of mission payloads (or payloads or PLs) are: scientific instruments, remote sensing instruments, navigation service transmitters, or communications equipment. A satellite may have one type of PL or a combination of payload types to accomplish its mission such as navigation, remote sensing, and communications. Shown below in **Figure 1** is a typical imaging satellite used for the remote sensing mission. Note the clear separation between the spacecraft bus that provides solar power and maneuvering capability via thruster, while the payload consisting of the camera and supporting communication devices such as antennas and guidance devices such as star trackers.

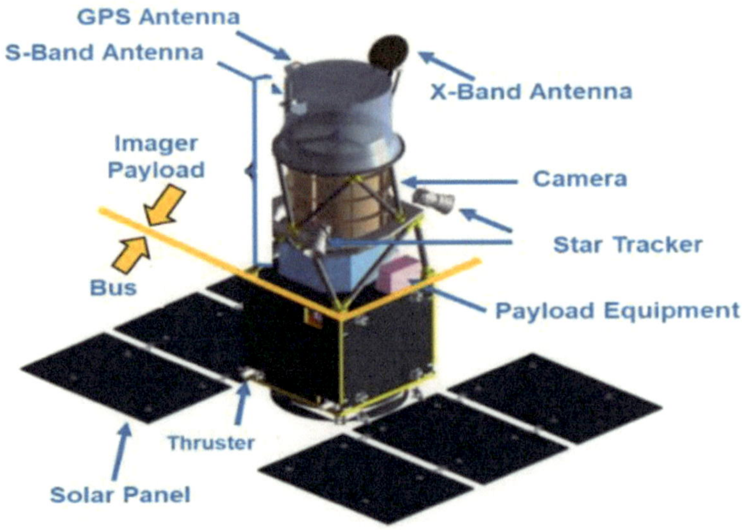

Figure 1.
A typical satellite with bus and payload separation.

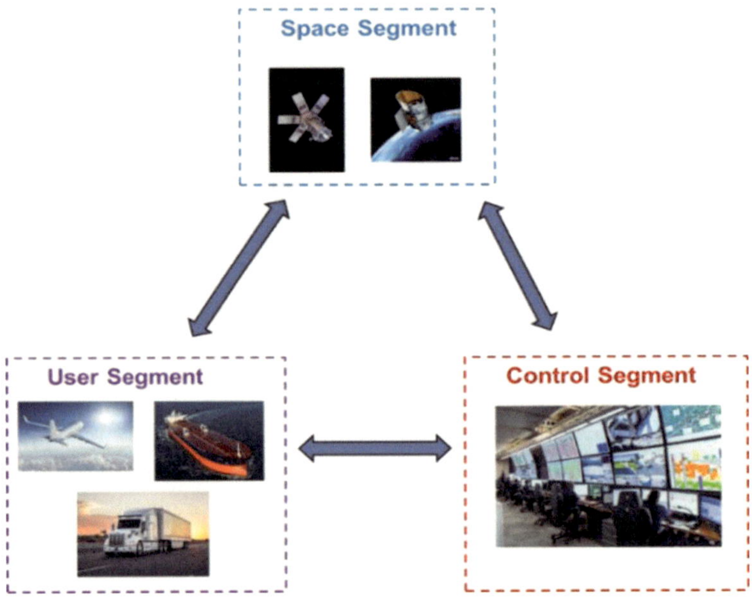

Figure 2.
The three main segments for satellite system.

Regardless of the mission type[1] and the payload that a spacecraft carries, a subsystem that must exist in all satellites is the communication subsystem that enables the spacecraft to communicate with the ground stations that control the satellite and to deliver the data that the mission requires. This chapter focuses on architecture and functionalities of the communications subsystem that usually resides on the satellite.

[1] Mission type has different meaning depending on context. For example, U.S. military satellite has three basic mission types: imaging/sensing; communications; positioning, navigation and timing (PNT) missions. For NASA space exploration, they are near-earth and deep-space missions.

There are three specific segments shown in **Figure 2** below that must work together for the larger overall system to provide communication, navigation, or any other type of missions:

- The space segment consisting of all satellites and associated equipment required for the mission applications and the launch vehicles used to deliver those satellites to orbit.

- The satellite control (or control) segment consisting of all the personnel, facilities, and equipment that are used to monitor and control all the assets in space. Practically, the control segment is also referred to as satellite ground segment because it is usually located on the ground.

- The user segment consisting of all the individuals and groups who use and benefit from the data and services provided by the payloads of the satellite and the equipment that allows this use.

2. Typical satellite major subsystems

In general, the space mission dictates the type of orbit[2], satellite design and its expected life cycle, and its operational scenarios. The PL design includes dimensions, interfaces, weight, physical characteristics, and basic utility needs (e.g., power consumption), which usually influences spacecraft (SC) bus design. The PL is often a unique and one-of-a-kind design tailored to meet specific mission requirements, frequently relying heavily on newer technology, while the satellite bus has the supporting function, and as such relies largely on existing or modified hardware such as batteries, inertial devices, and star trackers. Since PLs and their missions vary widely, so is this satellite bus supporting role.

Traditionally, the PL is considered a subsystem of the satellite bus that is designed to generally satisfy the corresponding mission requirements. The PL operational requirements sometimes impose specific requirements on the satellite bus that must be satisfied for the PL to accomplish its mission. This interdependence between satellite bus and PL subsystems has historically resulted in many nonstandard interfaces developed and implemented by the incumbent spacecraft builders. As a result, the aerospace industry has been moving toward a more standardized and commodity satellite bus framework that can potentially result in a tremendous cost saving approach.

As shown in **Figure 3** below, a satellite bus typically consists of the following subsystems: command and data handling subsystem (C&DHS); communications subsystem (CS); electrical power subsystem (EPS); propulsion subsystem (PS); thermal control subsystem (TCS); attitude control subsystem (ACS) also known as guidance, navigation and control (GNC) subsystem; structures and mechanics subsystem (S&MS); and life support subsystem for manned missions if required. The C&DHS will be described in detail below. The CS provides the satellite bus with the necessary communication functionalities to connect the user and ground segments to different satellite subsystems. The EPS provides the electrical power generation and distribution for various spacecraft subsystems. The PS provides maneuvers

[2] There are three main types of satellite orbits: low earth orbit (LEO) of 2000 km in altitude or less; geostationary (GEO) with altitude around 35,786 km; and medium earth orbit (MEO) with altitude between LEO and GEO.

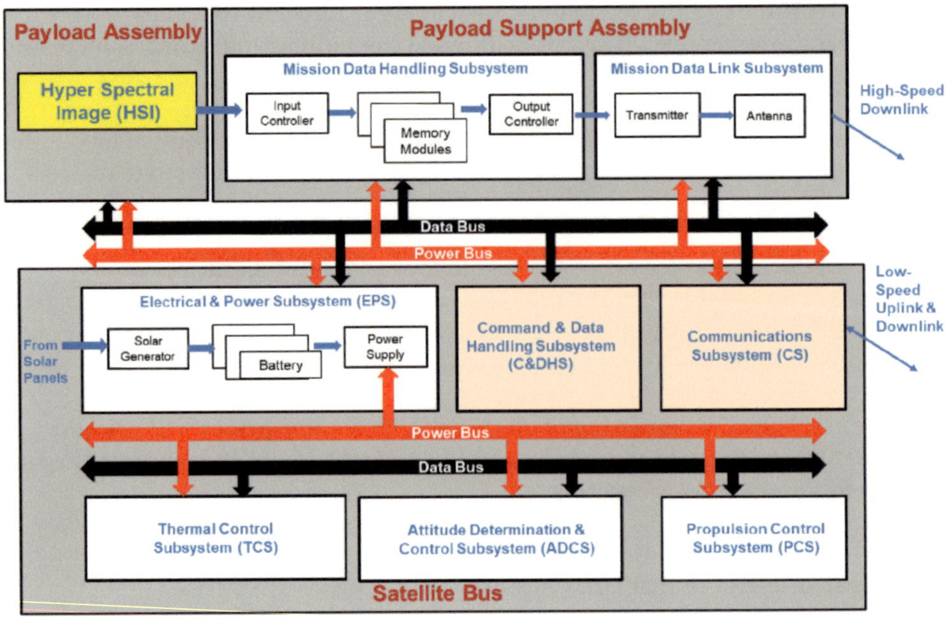

Figure 3.
A typical satellite bus and payload subsystem.

necessary for altitude, inclination adjustment, and momentum management adjustments. The TCS provides active thermal control from electrical heaters and actuators to control temperature ranges of equipment within specific ranges. The ACS provides proper pointing directions for the satellite subsystems, such as sun pointing for EPS to the solar arrays and earth pointing for CS. The S&MS provides the necessary mechanical structure to withstand launch loads by the launch vehicle, during orbital maneuvers, as well as loads imparted by entry into the atmosphere of earth or another planetary body.

On the other hand, a PL is tailored to a specific mission type. For example, a remote sensing satellite can have as its payload an electro-optical (EO) camera to take day-time pictures of the earth and then convert them to electrical signals that can be captured. Alternatively, the camera may also have infra-red (IR) sensors that enable the PL to see the earth at night, or microwave sensors that will let the PL "see" radio frequency (RF) signals from the earth at several radio frequencies

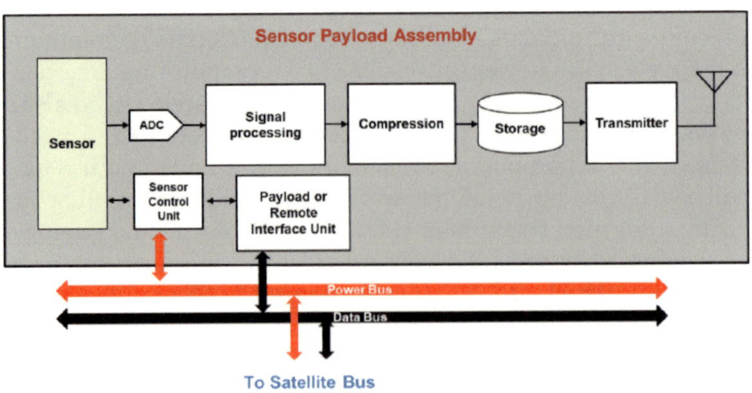

Figure 4.
A typical and generic sensor payload.

(RFs). These sensors can be classified as passive or active, and each of them can be further classified as imaging or sounding[3]. **Figure 4** below illustrates a generic imaging PL that will convert the sensor analog data into electrical signals that can be captured and transmitted to a ground station. Note the existence of a communication subsystem as part of this imaging payload.

3. Typical communications subsystems for satellite

In this section, the different typical modules of a satellite communication subsystem are discussed. In addition, the command and data handling subsystem, and command, telemetry and mission data processing subsystem will also be described in detail.

3.1 Antenna and RF front-end/back-end subsystems

At the physical layer, the communications subsystem starts with an antenna and the RF front-end transceiver. The antenna is the most important component of the communications subsystem where the electromagnetic (EM) signals are originated or received. The RF front-end/back-end is where the EM signal is being down/up-converted to baseband/RF signal to be demodulated/modulated for baseband signal recovery or downlink transmission, respectively. **Figure 5** below depicts a typical transmitter and receiver (transceiver) chain with the modulation and demodulation (MODEM), followed by the RF front-end and the antennas. The baseband communications function is carried out by the MODEM, whereas the RF portion is handled in the transceiver, RF front-end, and antenna sections.

Modulation is the name given to the process of impressing the wanted signal to be transported onto a radio frequency (RF) carrier, which is then conveyed over the satellite link and demodulated at the receiving terminal to extract the wanted signal from the carrier. Thus, modulation translates a baseband spectrum (at zero frequency) to a carrier spectrum (at RF range) and demodulation is the process of recovering the data at the receiver end of the link. Thus, the process requires a modulator

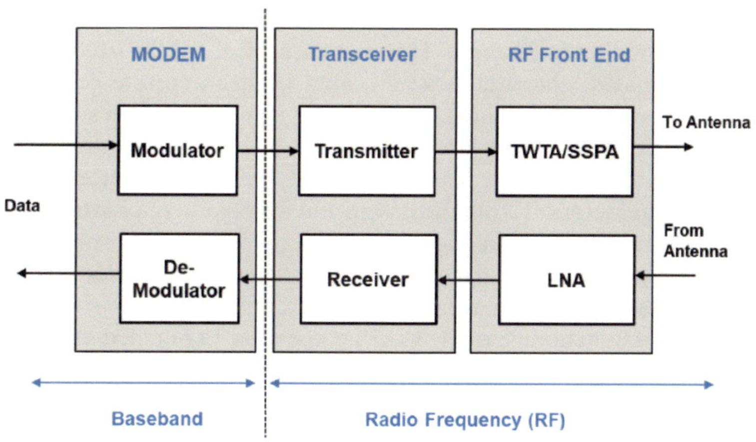

Figure 5.
Typical RF front-end chain.

[3] In this context, sounding means sending a radio signal and interpreting the results from the returned signals. Examples are radar and ranging signals.

and a demodulator, collectively known as a MODEM. The input to the modulator may require some initial processing such as filtering and amplitude limiting.

Before the RF signal is sent to the antenna, a traveling wave tube amplifier (TWTA) or solid-state power amplifier (SSPA) is needed to amplify the RF signal to a desired level for transmission. Conversely, after the RF signal is received by the antenna, a low noise amplifier (LNA) is needed to ensure that the received signal is brought up to the desired signal level with minimum noise before demodulation.

In addition to being lighter than TWTA, the achievable power efficiency for SSPAs is a major factor to support transmit phased arrays. Currently, the tube-based TWTA implementations are still the most cost-effective design, even though both options might be viable for lower power systems.

In increasing technical maturation over the years, the following types of spacecraft antennas have been used for satellite communications:

1. Low-gain omni and squinted-beam antennas for large earth coverage.

2. Increased gain types of satellite antennas (horn type and helix antennas) for medium earth coverage.

3. Parabolic reflectors, including multi-beam antennas with multiple feed systems for multiple user and small area coverage.

4. Deployable antennas, particularly to achieve more highly focused beams and support much high-gain multi-beam antennas.

5. Phased array feed and phased array antennas for scanning and hopping beams.

6. Optical communications systems, which have been used for intersatellite links and interplanetary communications, and increasingly being considered for earth-to-space systems.

In general, there are many different types of antennas, but the one most commonly associated with satellite communications is the parabolic dish antenna. These dish antennas have a narrow beam width, concentrating the energy of the radiated main beam into a smaller solid angle. This means more of the radiated energy reaches, or "illuminates," the satellite when using a dish antenna as compared to an omnidirectional, or "omni" for short, antenna. An example of dish antenna used on satellite is shown below in **Figure 6** for a Ku-band space to ground antenna (SGANT) mounted on the external stowage platform of the International Space Station (ISS).

There are several factors driving the design and development of satellite antennas. These include the need to reuse frequency bands because of limited spectrum allocations; the need to have antennas that can operate at higher frequencies with higher bandwidth; and the desire to deploy higher gain antennas at the same time minimizing the required size, weight, and power (SWAP) constrains. In practice, there are substantially more SWAP constrains for satellite antennas than on the ground stations, and this results in several design trade-offs between the space and control/user segments.

For example, the GEO orbit allows a high gain antenna to be pointed at a satellite with a minimum of tracking. Thus, a large dish can be used and remain virtually stationary without tracking a satellite as it moves around in its orbit. On the other hand, a low earth orbit (LEO) satellite that can cross from horizon to horizon in a few seconds can result in ground antenna installations that can be quite complex and expensive. Consequently, trade-offs need to be made to support the mission parameters of the whole satellite network.

Figure 6.
Example of a satellite dish antenna.

3.2 Command and data handling subsystem

The term "command and data handling subsystem" (C&DHS) was referred to as "On-board Computer" (OBC), which is a legacy of the past in which many satellite functions were performed by analog circuits with the help of an OBC. With the current shift toward the digital domain, the term OBC does not fully cover the topic anymore thus C&DHS is being used instead. An appropriate analogy to describe the C&DHS subsystem is to regard it as the brain and nervous system of the spacecraft.

The function of a C&DHS subsystem is to perform onboard processing and operations and internal communication [3, 4]. The task of managing the operations of the spacecraft subsystems is nowadays performed mostly by software in an autonomous manner and is generally categorized as onboard operations. The software is also responsible for preparing the data to be downlinked and handling any commands that are received from satellite operators on the ground. Lastly, the C&DHS facilitates and controls all internal communications (consisting of commands, telemetry, and tracking data) between the different satellite subsystems. The basic functions of the C&DHS can be summarized below:

1. Receives commands from the command or user segment through the telemetry, tracking, and control (TT&C) subsystem.

2. Decodes, executes, and/or distributes those commands to/from the onboard computer.

3. Collects and formats telemetry data from all space vehicle (SV) units.

4. Distributes telemetry for downlinking. Provides a platform for bus flight software (FSW).

5. Additional functions include ranging processing for satellite tracking purpose, satellite timekeeping, computer health monitoring (watchdog), and security interfaces.

An overview of the architecture of C&DHS in a typical satellite is provided in **Figure 7** below. In this figure, all components are connected to each other via a

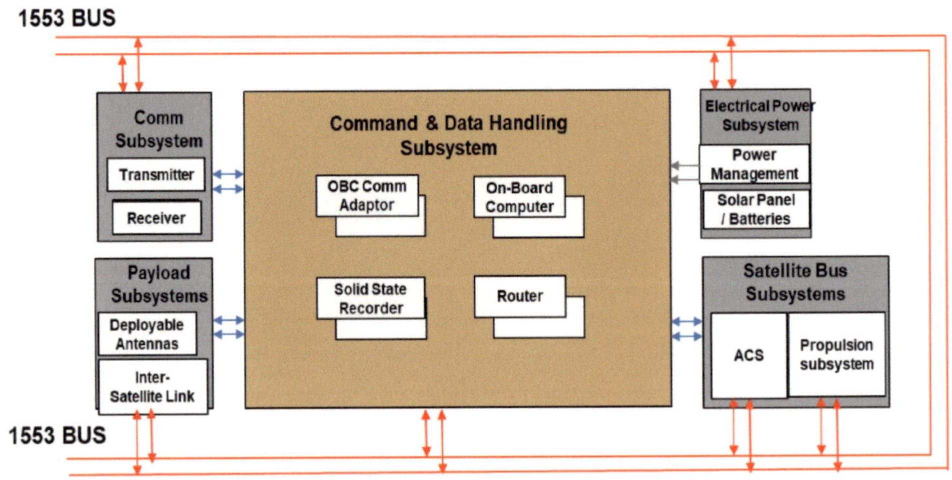

Figure 7.
Block diagram of a typical command and data handling subsystem.

common low-speed data bus in red color, typically compliant with MIL-STD 1553 or other standards. Also shown is the data connection in blue from the C&DHS to other components, which is more customized and high-speed in nature depending on the design.

The heart of the system is the C&DHS' onboard computer (or OBC) that runs the software responsible for managing the onboard operations. The OBC is tightly linked to the electrical power subsystem (EPS). The main reason is the importance of the available and consumed power for managing onboard spacecraft operations. For instance, by continuously querying the EPS on the available power, the OBC can decide to turn off non-critical subsystems to prevent vital systems from shutting down from lack of power. Secondly, the OBC must be able to command the EPS to disable or enable different subsystems throughout the various phases of the mission. Since the amount of transmitted data between these two subsystems is small, a low-speed data link is sufficient, although there is a new trend to incorporate high-speed standard link such as SpaceWire[4] to satisfy increasing demand for data volume.

The OBC is also responsible for receiving, interpreting, and executing commands from ground operators via the radio receiver. Using low-speed radio transmitters, the OBC also sends packets of housekeeping data, or telemetry, to the ground station. The purpose of the housekeeping data is to give the operators on the ground an overview of the spacecraft health and its general condition.

Some small satellites only have a single low-speed transmitter, so the housekeeping and payload data are combined over the same link. For larger satellites with payloads capable of producing vast amounts of data, a dedicated high-speed data link is used to store the data on an onboard storage system. When the satellites pass over a ground station, the OBC commands the high-speed radio transmitter to retrieve and transmit the previously stored payload data through another dedicated high-speed link from the onboard storage system. This approach frees the OBC from having to process large amounts of data and allows it to devote its internal

[4] SpaceWire is a spacecraft communication network based in part on the IEEE 1355 standard of communications. It is coordinated by the European Space Agency (ESA) in collaboration with international space agencies including NASA, JAXA, and RKA

resources for time critical operations and communicates with the PL and all other subsystems through the low-speed data links. This would include the requirements to retrieve information on the health, perform critical interventions as well as to command these subsystems to perform various actions according to the operational arrangement of the mission.

3.3 Typical command and telemetry processing subsystem

The telemetry, tracking, and control (TT&C) subsystem of a satellite provides a connection between the satellite (space segment) and the ground facilities (control or user segment). The purpose of the TT&C function is to ensure the satellite performs correctly. As part of the satellite bus, the TT&C subsystem is required for all satellites regardless of the mission type. The TT&C subsystem has three specific tasks that must be performed to ensure a successful mission:

1. Telemetry: the collection, processing of health, and status data of all spacecraft subsystems, and the transmission of these data to the control segment on the ground. This requires not only a telemetry system on the spacecraft but also a global network of ground stations around the world, unless the satellite space network includes intersatellite links that can relay the data to designated satellite and downlink to the appropriate ground station. **Figure 8** below illustrates the processing of telemetry data by the C&DHS. Here the different health information and status information sent from various subsystems are collected by the telemetry input interface, fed to the C&DHS processor, buffered, encrypted, and sent down to the ground station.

2. Tracking: the determination of the satellite's exact location by the control segment and where it is going via the reception, processing, and transmitting of ranging signals. This requires a ranging system on the spacecraft and a data collection ground network for this tracking function to work.

3. Command and control: the reception and processing of commands for continuous operation of the satellite. Usually a ground system is required, although advanced spacecraft designs have evolved toward "autonomous operations" so that many of the control functions can be automated onboard and do not require ground intervention except under emergency conditions. A typical command processing scenario is illustrated in **Figure 9** where serial command bit stream from the command receiver is received by the command input interface, where the relevant commands are extracted and sent to the appropriate subsystems via a serial or parallel interface.

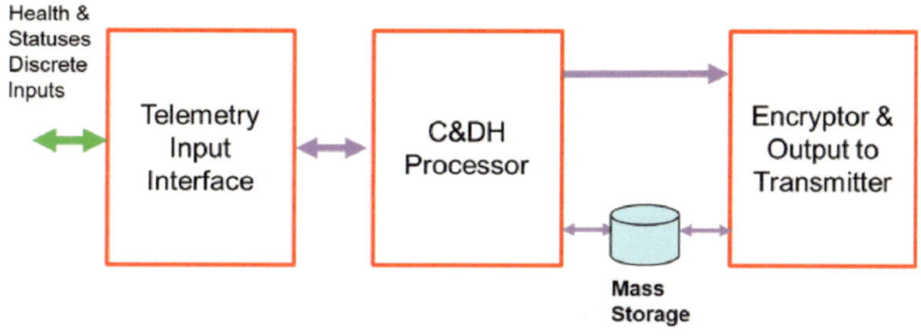

Figure 8.
Telemetry processing by C&DHS.

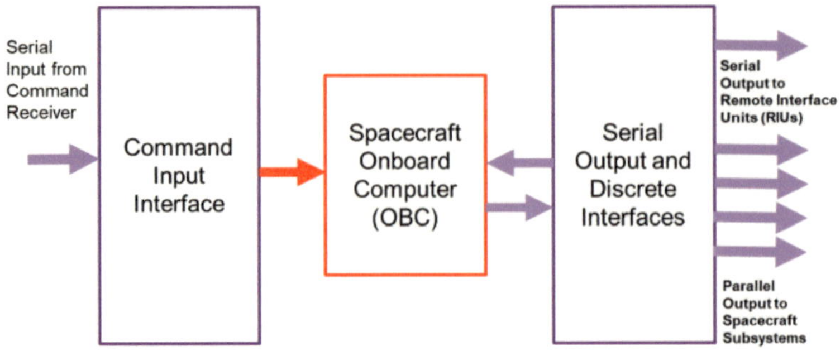

Figure 9.
Command and control message processing by C&DHS.

3.4 Typical mission data processing subsystem for communication applications

For communications payload, the onboard switching systems are designed to make more efficient use of a satellite communication network, especially those that employ multi-beam technology that entails onboard switching to interconnect uplink and downlink beams with a high degree of efficiency.

Figure 10 below summarizes the functional block diagram of a channelized transponder processor assuming a digital implementation of the channelized transponder filtering and switching function. Any signal within the receiver bandwidth is down-converted to an intermediate frequency (IF) or baseband and digitally sampled. These samples are digitally filtered, stored, and routed to the switch port corresponding to the desired downlink beam. This routing is achieved by a simple readdressing of the stored digital samples within a common output buffer memory or by a more traditional digital switch implementation.

For most sensing payload and as shown in **Figure 4** above, the sensor analog data are collected onboard, digitized, buffered if necessary, and transmitted down to ground station for processing. This is due to the complexity of sensing mission

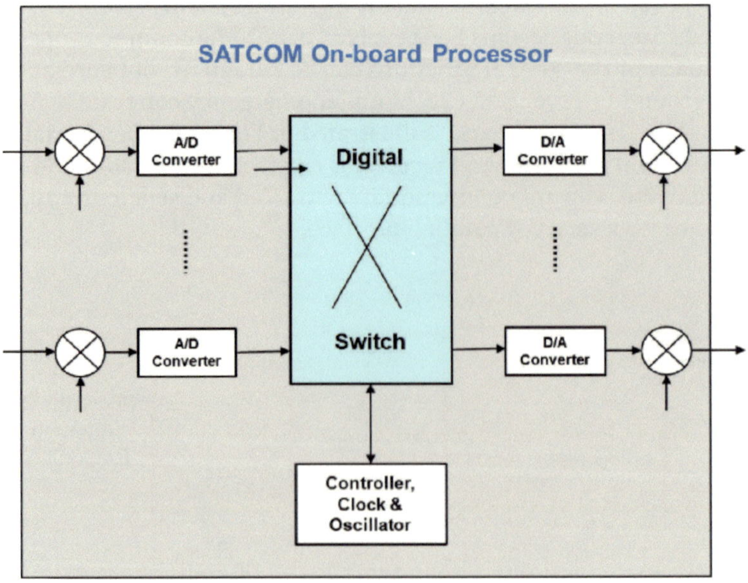

Figure 10.
Channelized processor for communications payload.

data processing and the lack of onboard computational power to accomplish these tasks. An example of onboard PL processing for passive electro-optical (EO) remote sensing is shown in **Figure 11** below, where the reflected light from earth is passing through a combination of optical lenses and charge coupled device[5] (CCD) whose output is an analog signal that would be conditioned by analog filters before being digitized, compressed, and sent down via a mission data downlink to the ground station for processing. There, the data are decompressed, and image is enhanced by appropriate algorithms and displayed for users.

Typical data volume collected by sensing payload is large, and peak rates can produce data at much higher speeds than TT&C; thus, a separate downlink for mission data is needed. Depending on the system, this mission data downlink to a ground station can either be performed using a dedicated mission direct downlink, or indirectly via a relay broadband communications satellite. Sensing satellite can be positioned in GEO, MEO, or LEO orbits, and can have many possible mission data downlink architectures based on mission requirements. For example, a LEO sensing satellite can either buffer its mission data until within view of a dedicated ground station for downlink, or it can forward its mission data to a relay satellite that can ensure that the mission data can be downlinked to a designated ground station.

Another example of active remote sensing is a synthetic aperture radar (SAR) mission, where returned radar signals are collected onboard and sent to the ground to be correlated and form an image of the ground surface. This type of remote sensing does not heavily depend on sun light and other weather affects. Applications for SAR include agriculture, geology, geohazards, ice, oil spills, and flood monitoring. Several emerging applications such as forestry, ship detection, and others are possible [1]. An example of a SAR mission is the NASA-ISRO Synthetic Aperture Radar (NISAR) [5], which is a collaborative earth-science mission between NASA and the Indian

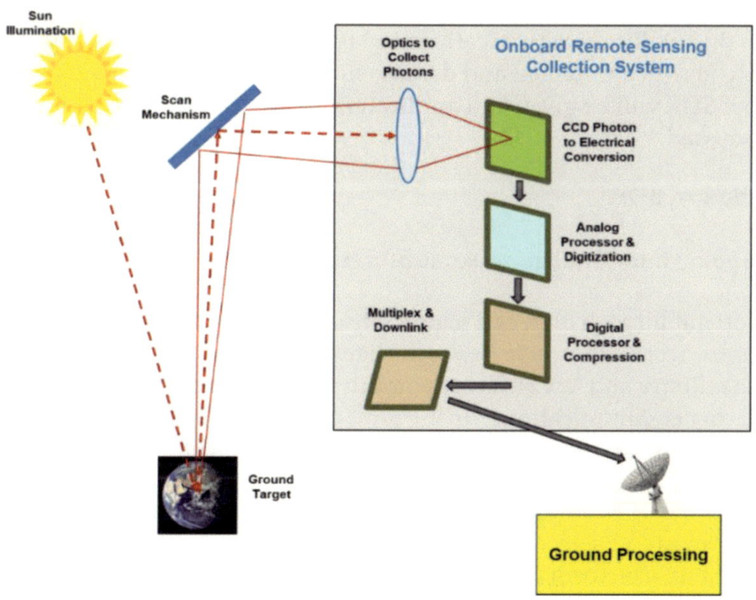

Figure 11.
Onboard image processing for an EO application.

[5] A CCD is an integrated circuit etched onto a silicon surface to form light-sensitive elements represented by what are called pixels. Photons incident on this surface generate a charge that can be read by electronics as a voltage and turned into a digital copy of the light patterns falling on the device.

Space Research Organization (ISRO). The sensing payload features an L-band SAR instrument and an S-band SAR instrument. The simultaneous dual-frequency radar system at peak rates will produce data at gigabit-per second speeds, which drives the data-volume requirements at a minimum of 35 Terabits per day of radar science data to the ground. This is a direct mission downlink system with three designated ground stations. The payload communication system uses a 70-cm high-gain antenna with two synchronized transmitters in a dual-polarization configuration with each transmitter providing 2.4 Gbps of coded data with an aggregate rate of 4.8 Gbps.

4. Software defined radio (SDR) and applications to satellite

Traditional communications systems are designed for and constrained to a specific waveform(s) operating over predetermined frequencies, bandwidths, and signal modulation types. This paradigm works well when the requirements and constraints of the communication link and network protocol are well understood prior to design.

As a result, most radios in today's world have very dedicated uses. A car key fob is designed only to unlock or lock your car door, while a smart phone radio connects to the Internet through various wireless communication protocols. Although these examples vary in complexity of the hardware, they both cannot operate outside the confines of their physical layer implementation. Consequently, RF hardware with a narrow focus is not suitable for applications with a broader communication scope.

A single software defined radio (SDR) with a flexible RF front-end combined with modern computing power can be used for the above applications plus more. In addition, a radio with a flexible hardware and software architecture can also lead to more innovation in the communications industry. Because of the rapid development nature of software, an engineer or researcher can experiment with novel ideas and SDR waveforms that would not be achievable with a traditional radio.

SDR in the satellite communications industry has become a growing trend, particularly in the commercial and defense industries. In the following section, an overview of SDR will be given and applications of SDR in satellite communications will be discussed.

4.1 Overview of SDR

Before going into SDR basics, some of the SDR advantages are [6]:

- Interoperability: an SDR can seamlessly communicate with incompatible radios, or work as a bridge between them. For example, different branches of the military and law enforcement can use many incompatible radios, thus hindering communications during joint operations. A single multichannel SDR can work with all these different radios and provide interoperability.

- Efficient use of resources under varying conditions: for example, a low-power waveform can be selected if the radio is running low on battery, while a high-throughput waveform can be used to quickly download a file. This flexibility is one of the first reasons why SDR became popular.

- Opportunistic frequency reuse in SDR using cognitive radio[6] (CR) technology: if the "owner" (or primary user) of a spectrum band is not using it,

[6] A cognitive radio (CR) is a radio that can be programmed and configured dynamically to use the best wireless channels in its vicinity to avoid user interference and congestion.

an SDR-CR can "borrow" the spectrum until the owner comes back. This technique has the potential to dramatically increase efficient use of radio frequency spectrum.

- Reduced obsolescence: an SDR can be field upgraded to support the latest communications standards. This capability is especially important to radio with long life cycles such as those in satellite communications.

- Lower cost: a single SDR can be adapted for use in multiple markets and for multiple applications. For example, a single radio can be sold to cell phone and automobile manufacturers to significantly reduce cost.

- Research and development: SDR can be used to implement many different advanced waveforms, e.g., code division multiplexing access (CDMA) or orthogonal frequency division multiplexing (OFDM), for real-time performance analysis. Performance studies can be conducted much faster and often with higher fidelity than simulations.

On the other hand, some of the disadvantages for SDR are:

- Cost is the most common argument against SDR. A single key fob is based on a very inexpensive ASIC[7]; however SDR is heavily reliant on FPGA,[8] which is much more expensive. This is even more significant for high-volume, low-margin consumer products.

- The second most common argument against SDR is increased power consumption with increased DSP complexity and higher mixed-signal/RF bandwidth. Power consumption in an FPGA or GPP for flexible signal processing can easily be 10 times higher than in ASIC. Also, wideband analog-to-digital converters (ADCs), digital-to-analog converters (DACs), and RF front-ends consume more power than their narrowband equivalents.

- Increased time and cost to implement the radio: it can take much more engineering effort to develop software/firmware for multiple waveforms than for one, especially if it must be compliant with a military standard such as JTRS[9].

- Changing specifications and requirements: this usually happens when the SDR design must support not only a set of baseline waveforms but also anticipate additional waveforms.

- Increased schedule risks: since SDR is still a relatively new technology, it is more difficult to anticipate schedule problems. Also, it is difficult to thoroughly test the radio in all the supported and anticipated modes.

- Limited technical scope: SDR only addresses the physical layer and will require cooperation from upper layers for throughput improvements.

---

[7] Application-specific integrated circuit (ASIC) is designed for specific purpose.

[8] Field Programmable Gate Array (FPGA) is an integrated circuit designed to be configured by a designer after manufacturing thus much more general purpose.

[9] The Joint Tactical Radio System (JTRS) aims to replace existing radios in the U.S. military with a single SDR to enable new frequencies and waveforms added via software or firmware upload.

4.1.1 SDR basics

The general definition for a SDR is *a radio with some or all its physical layer behavior defined through means of software* [7, 8]. SDRs are incredibly valuable devices as they allow the end user the ability to traverse the RF spectrum at variable sampling rates. The fundamental qualities that make up an SDR are the flexible specifications and the ability to transform the analog signal using digital signal processing (DSP).

A radio can be categorically separated into receivers and transmitters. For this section, the receiver implementation will be considered as it is generally more interesting and complex. A block diagram of an SDR receiver is shown below in **Figure 12**. The following sections will present the anatomy of the SDR that differentiates it from a traditionally designed radio.

4.1.2 RF front-end

The purpose of the RF front-end (RFFE) is to isolate the desired signal received by the antenna from interference signals. To achieve this, the signal of interest must be brought down to lower frequency for digital conversion while mitigating the side effects from filtering during the frequency conversion process. A flexible RFFE for SDR must be designed so that the frequency and bandwidth are controllable by software. Depending on the system requirements and the available RF component specifications, there are several ways to achieve this.

One of the most common RFFE designs for analog radios is the heterodyne receiver. A heterodyne receiver, shown in **Figure 13** below, works by mixing down the received signal from its carrier frequency to a lower intermediate frequency (IF). The signal at IF can now be more conveniently filtered, amplified, and processed. A super-heterodyne receiver uses a fixed IF that is lower than the carrier frequency but higher than the signal bandwidth and often uses two stages of down conversion to reduce the filtering requirements at each stage.

Another popular RF front-end architecture generally used for low-power applications is called zero-IF. A zero-IF receiver, shown in **Figure 14** below, uses a single mixing stage with the local oscillator (LO) set directly to the desired carrier frequency to convert directly to baseband in-phase and quadrature signals. Because mixers tend to have high power consumption and only low-pass filters are required, the simpler zero-IF provides improved power efficiency over a heterodyne architecture. However, the zero-IF implementation is more susceptible to IQ imbalances of the in-phase and quadrature oscillators, which will produce anomalies in the signal constellation. LO leakage may also self-mix through the RF ports creating a large DC bias. Both issues can be corrected using digital signal processing.

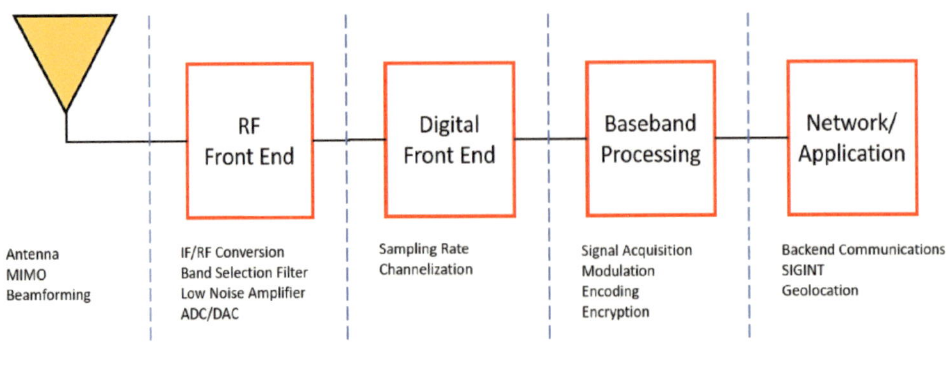

Figure 12.
A block diagram of an SDR.

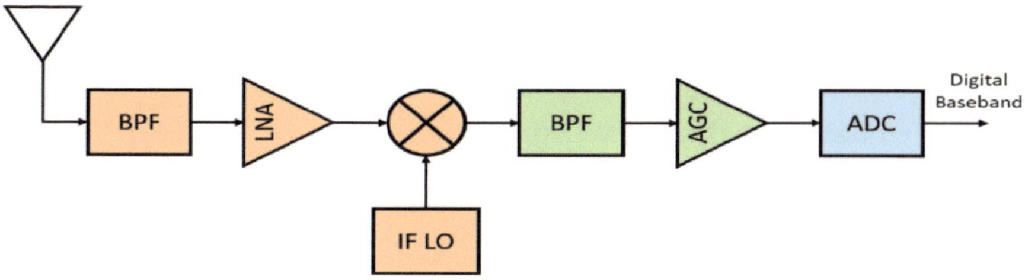

Figure 13.
Heterodyne receiver.

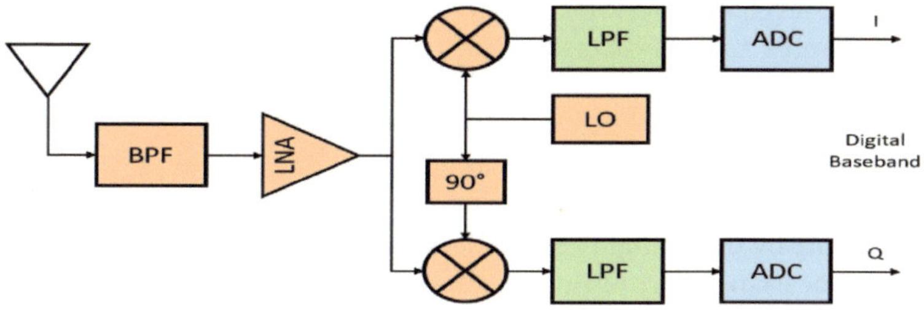

Figure 14.
Zero-IF receiver.

The analog-to-digital converter (ADC) is responsible for converting a continuous-time signal to a discrete-time one. To translate signals from the analog to digital domain, an ADC must perform two fundamental steps: sampling and quantization. Sampling is the process of reading voltages at discrete-time intervals. Quantization is the process of converting these voltage readings into binary outputs. ADC performance can be evaluated based on various parameters, such as: signal-to-noise ratio (SNR), dynamic range, bit resolution, sampling rate, and power dissipation. The ADC dictates the DSP limitations of the SDR. Generally, the sampling rate should be at least twice the desired bandwidth of your signal. The ADC should be chosen to match the capability of your processor and specifications of the signals of interest.

4.1.3 Digital front-end (DFE)

The two main functions of a digital front-end are sample rate conversion (SRC) and channelization. Once a signal has become digitally converted, the samples need to be further primed for digital processing. Operating the ADC at a fixed rate simplifies its clock generation; however, it may be necessary to convert the sampling rate to match the sampling rate required to demodulate certain waveforms. Most wireless signals generally operate with specific symbol or chip rates that are specified by their respective standard. Depending on the RFFE design and signal type, channelization may be required to select the channel of interest.

SRC represents a classic sampling theorem problem. Converting sampling rates can introduce undesirable effects such as aliasing, an effect that causes frequency components to overlap. SRC can be achieved digitally through the processes of decimation and interpolation. To mitigate aliasing, decimation is performed by using an anti-aliasing filter followed by subsampling, which is essentially removing samples at certain intervals. Interpolation is a method of calculating values to add values in between samples. Channelization works by using digital down conversion, the process of digitally mixing down a signal to baseband with a numerically controlled oscillator.

4.1.4 Digital signal processing

SDRs have an array of devices to choose from for the required DSP application, each with their own strengths and weaknesses. An SDR may integrate multiple processor types and partition the signal processing chain to optimize each processor. The following criteria should be considered when evaluating the various processor types: flexibility, modularity, and performance. The three digital hardware choices this section will consider are the general-purpose processor (GPP), digital signal processor (DSP), and the field programmable gate array (FPGA).

A GPP is the typical microprocessor designed to handle a wide variety of generic tasks that can be found in your everyday personal computer. They are generally designed to have large instruction sets and highly capable of implementing and performing complex arithmetic tasks such as modulation/demodulation, filtering, fixed/floating point math, and encoding/decoding. Some commonly used GPP architectures are x86/64 and Advanced RISC Machine (ARM). The advantage of using a GPP is the wide availability, flexibility, and ease of programmability. Several GPP-based SDRs, such as Universal Software Radio Peripheral (USRP) and the LimeSDR, operate by digitizing the baseband signal and performing the required digital signal processing on computers. These types of SDRs are popular among university researchers and hobbyists due to the relative ease of obtaining and developing their applications.

Because the GPP was designed with such a broad focus, latency, speed, and power efficiency may be a limiting factor depending on the application. Many wireless communication standards have strict real-time and large processing bandwidth requirements that most modern CPUs cannot meet due to processor architecture and operating system design. .

A DSP is a microprocessor optimized for digital signal processing applications with the ability to be programmed with high-level languages. Although a GPP can contain much of the same functionality, the DSP performs the same digital signal processing operations more quickly and efficiently due to its reduced instruction set computer (RISC) architecture and parallel processing. The reduced instruction set limits the essentials but contains optimizations for common DSP operations such as multiply accumulate (MAC), filtering, matrix operations, and fast Fourier transform (FFT). DSPs are commonly sold in two variants: optimized for power efficiency and optimized for performance; and are used in applications such as base stations and edge devices. Power consumption is also minimized by reducing the silicon footprint that would be in GPPs sophisticated cache and peripheral subsystems.

Although DSPs have been commonly deployed in the past decades, they serve as a middle ground between GPPs and FPGAs with regard to flexibility, performance and efficiency. Field-programmable gate array (FPGA) offers more parallelism, higher data rates, and better power efficiency than DSP, but is not well suited for control applications, such as implementing the network/protocol stack. This is due to the limited amount of memory in FPGA and for this reason it is often paired with GPP.

A FPGA is an array of programmable hardware logic blocks, such as general logic, memory, and multiplier blocks, that are wired together via a reconfigurable interconnect to generate an integrated circuit for several designs with the ability to quickly switch between configurations. FPGA configurations are programmed using hardware description language (HDL), which is also used for ASIC. Because a FPGA functionality is defined at the hardware level and can be implemented using parallelism, it can perform DSP algorithms at much higher rates than DSPs and GPPs. FPGA consumes more power and requires more space than ASICs but provides more programmability and flexibility than ASIC. A big consideration for using FPGAs for SDR is the domain knowledge requirement for developers. Developing on FPGAs can be time consuming and require an extensive understanding of the target hardware architecture.

When the system requirements exceed the capabilities of a singular processor type, a comprehensive solution may include a combination of the above processor types. A common processing architecture in the defense industry comprises of a FPGA, DSP, and GPP. In this paradigm, the FPGA is responsible for high data rate signal processing tasks, such as sampling and filtering, the DSP handles demodulation and protocol, and the GPP performs control-related tasks, such as the user interface and algorithmic processing. Implementing such a system can become a complex management task to coordinate the processing flow; however, the system can benefit greatly by optimizing overall performance based on the strength of each processor.

4.2 Applications of SDR to satellites

For space applications, SDR has unique challenges such as extreme radiation and temperature environment, autonomous operational requirements, limitations on size, weight and power (SWAP), and the need for reduced development time and increased reliability in agile prototyping. In this section, recent applications of software defined radio to satellite, as well as the current status of radiation-hardened SDR components, are presented.

4.2.1 NASA STRS

Recognizing early on that a standard and open architecture is needed to encourage reuse and portability of software, NASA developed an open architecture specification for space and ground SDRs called the Space Telecommunications Radio System (STRS) [9]. From this standard, several compliant systems have been built and demonstrated in radios on the International Space Station (ISS) and several ground stations. It was also the intention of NASA that the STRS architecture should be used as baseline for many future NASA space communications technologies.

In a nutshell, the STRS standard consists of hardware, configurable hardware design, and software architectures with accompanying description, guidance, and requirements. The three main hardware functionalities are connected by the Hardware Interface Description[10] (HID) and described and shown in **Figure 15** below:

1. General processing module (GPM) consists of the general-purpose processor; appropriate memory; spacecraft bus (e.g., MILSTD-1553, Space Wire); interconnection bus (e.g., PCI); and the components to support the configuration of the radio.

2. Signal processing module (SPM) where signal processing is used to handle the transformation of digital signals into data packets. Its components include ASICs, FPGAs, DSPs, memory, and connection fabric/bus (e.g., PCI, flex-fabric).

3. RF module (RFM) handles the RF functionality to transmit/receive the appropriate digital signal. Its components include RF switches, digital-to-analog converter (DAC), analog-to-digital converter (ADC), diplexer, filters, low-noise amplifiers (LNAs), and power amplifiers (PAs).

[10] In a HID, the hardware architecture requirements are written so that the hardware provider defines the functional modules of the system and publishes the functions and interfaces for each module and for the entire STRS platform [Wikipedia].

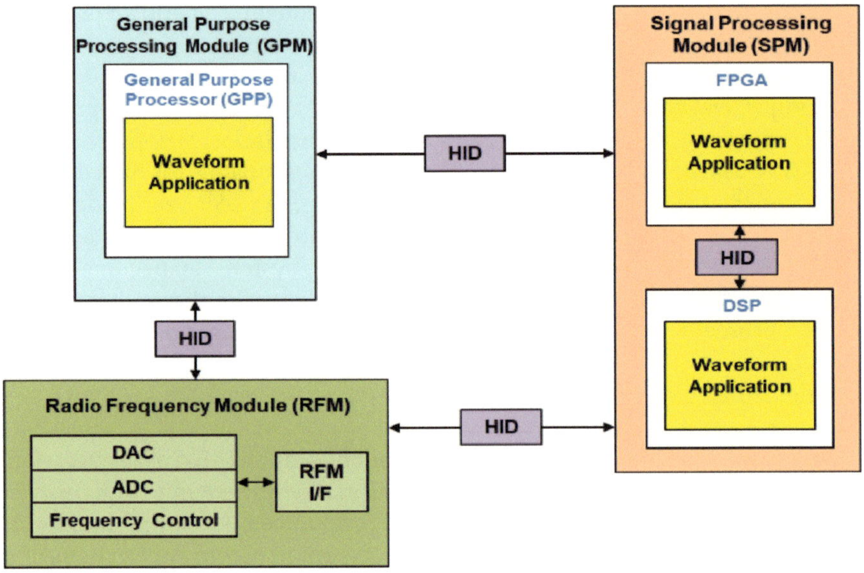

Figure 15.
NASA STRS' three main hardware functionalities.

In STRS terminology, software includes source code, object code, executables, etc. implemented on a processor. As shown in **Figure 16**, the STRS software architecture uses three primary interfaces: the STRS APIs, STRS hardware abstraction layer[11] (HAL) specification, and the Portable Operating System Interface[12] (POSIX®). The STRS APIs provide the interfaces that allow applications to be instantiated and use platform services.

Configurable hardware designs are the items and designs, such as hardware description language (HDL) source, loadable files, data tables, etc., implemented in a configurable hardware device such as a FPGA.

STRS encourages the development of applications that are modular, portable, reconfigurable, and reusable. The STRS software, configurable hardware design, metadata, documentation for STRS applications, STRS devices, and operating environments (OEs) are submitted to NASA STRS Application Repository to allow applications to be reused in the future with appropriate release agreements.

4.2.2 SDR applications in CubeSats

CubeSats[13] are increasingly popular spacecraft platforms for mission-oriented experiments that can be accomplished via quick prototyping and launches [10–12].

---

[11] Hardware abstraction layer (HAL) is a layer of programming that allows a computer operating system to interact with a hardware device at a general or abstract level rather than at a detailed hardware level [Wikipedia].

[12] The Portable Operating System Interface (POSIX) is a family of standards specified by the IEEE Computer Society for maintaining compatibility between operating systems. POSIX defines the application programming interface (API) for software compatibility with variants of Unix and other operating systems [Wikipedia].

[13] CubeSats are a class of Small Satellites (SmallSats) weighing between 1 kg and 10 kgs that use standard size and form factor of 1 U (one unit) of 10 cm x 10 cm x 10 cm. A standard 3 U CubeSat (10 cm x 10 cm x 34 cm, 5 and 6 kg) has been demonstrated to support real mission, and larger CubeSats (6 U, 12 U, and 24 U) are developed.

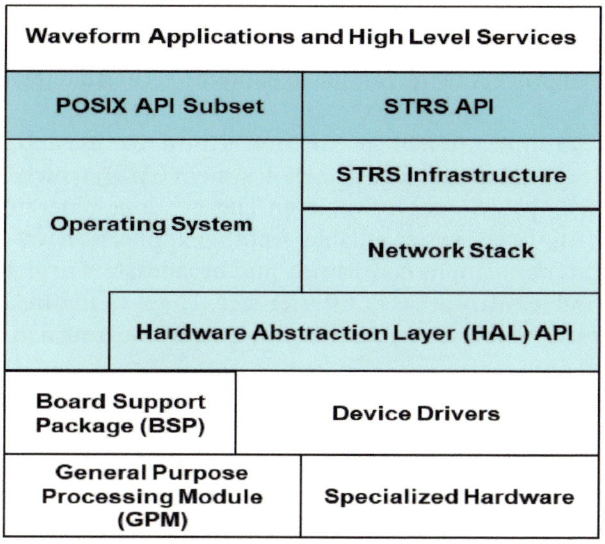

Figure 16.
STRS software architecture layers.

This short development timeline is due to the use of commercial-off-the-shelf (COTS) technology that typically has limited resilience to the space environment. Therefore, CubeSat usage has largely been limited to experiments or applications where high availability is not the main objective.

In general, SDR technology will allow for on-orbit flexibility via reconfigurability of data management, protocols, multiple access methods, waveforms, and data protection. SDR processing requirements are inherently scaled to the application. The availability of modular, high-performance sequential and parallel processors that are resilient to radiation upsets allows the tailoring of hardware architectures to the application and to the CubeSat platform. This is especially suitable for missions that require the flexibility to support multiple TT&C and mission data from multiple satellites and ground stations [13–15].

Given the provided mission flexibility, implementing an SDR on a CubeSat could significantly increase the required processing capacity and thus the size, weight, power and cost (SWAP-C) of the SDR implementation. Consequently, most current CubeSat SDR design and implementation are still customized depending on the mission requirements. In [16], some of the current COTS SDR hardware and software platforms such as GomSpace, Ettus Research USRP, EPIQ Solutions, Lime Microsystems, FunCube, and RTL SDR are described and categorized in decreasing cost and mass to illustrate the heterogeneous nature of SDR in CubeSat applications. Also described are a number of space and ground segment systems built to be (or have been) launched using these COTS SDRs or components thereof. What would be needed is a standard for CubeSat SDR similar to NASA STRS to ensure that hardware and software reuse can be incorporated into future CubeSat developments.

4.2.3 SDR applications in emitter location from space

A pioneering commercial application of SDR in space is the HawkEye 360 (HE360) system [17] that was launched on 3 December 2018. HE360 system consists of three identical spacecrafts and their primary payload is a SDR with custom RF front-end along with VHF Ku-band antennas. This Pathfinder mission[14] was to enable onboard reception and geolocation of different types of terrestrial RF signals using signal processing technique to combine received data from all three payloads[15].

One commercial application of this mission is the detection and geolocation of a maritime vessel's automatic identification system (AIS), which broadcasts the locations generated by GPS-enabled receiver. The locations generated by AIS can be disabled or spoofed, therefore not reliable. Another application would be to allow regulators, telecommunications companies, and broadcasters to globally monitor spectrum usage and identify areas of interference. The system can also be used to help large area search and rescue operations by quickly locating activated emergency beacons.

The SDR developed for the Pathfinder payload consists of an embedded processor system and three baseband processors. The baseband processor was built around the Analog Devices 9361 (AD9361) System on Chip (SoC) product, which is a highly integrated RF transceiver that combines high-speed ADCs and DACs, RF amplifiers, filtering, switching plus more. The HE360 payload supported up to three receiver channels (one AD9361 per channel) that can be simultaneously processed on separate frequencies. In addition, the signal processing subsystem takes advantage of open-source software and firmware code to allow system development to proceed without knowing the final space hardware. GNURadio[16] was selected for being a free and open-source toolkit for SDR and widely used in small space projects for ground software processing.

4.2.4 Radiation hardening and its current status on SDR hardware

In space, most semiconductor electronic components are susceptible to radiation damage, thus radiation-hardened (or rad-hard) components are required and normally developed based on their COTS equivalents with variations in design and manufacturing[17] to reduce the susceptibility to radiation. Consequently, rad-hard components tend to lag behind most recent COTS developments. Depending on mission requirements, rad-hard products are typically selected and tested using popular metrics such as total ionizing dose[18] (TID), and single event effects[19] (SEEs).

---

[14] A Pathfinder mission is usually a demonstration to prove that a system can successfully achieve a specific objective before a full mission can deploy.

[15] By comparing time-of-arrival (TOA) and frequency-of-arrival (FOA) measurements between pairs of receivers, the position of a signal can be computed.

[16] GNU is a recursive acronym for "GNUs Not Unix!" chosen because GNU's design is Unix-like but differs from Unix by being free software and containing no Unix code.

[17] Design trade-offs choosing more better radiation tolerance integrated circuit technology, low-power Schottky vs. Emitter Couple Logic (ECL) vs. bipolar over CMOS. Rad-hard chips are often manufactured on insulating substrates (silicon on insulator, silicon on sapphire, silicon carbide, gallium nitride) instead of the usual semiconductor wafers. Shielding the package against radioactivity to reduce exposure of the bare device.

[18] TID causes slow gradual degradation of the device's performance and is measured in rads.

[19] SEEs are caused by a single energy particle and can be (a) non-destructive Single Event Upset (SEU) causing transient pulses in logic or support circuitry or as bitflips in memory cells or registers, (b) potentially destructive Single Event Latch-up (SEL) that may be cleared by a power reset, and (c) destructive Single Event Burnout (SEB) or Single Event Gate Rupture (SEGR), which is irreversible.

Per US DoD MIL-PRF-38535 J standard [18], an ideal integrated circuit for space applications is the qualified manufacturing line[20] (QML) Class V with radiation hardness assurance[21] (RHA) level identified in the part specification. From the perspective of payload designer and developer, only Class V is space quality and should be the main factor for selecting SDR hardware components.

The FPGA is perhaps the most important component of an SDR and has a long history for manufactured QML class V parts where rad-hard Xilinx and Actel (now Microsemi) FPGAs were studied [19]. Currently, Xilinx is the major player for space-qualified QML level V products used in actual payloads with many more devices under development. The rad-hard DSP products also follow the QML process, with Texas Instrument (TI) currently taking the lead for in-flight payloads with many offerings in space-qualified RF components in addition to DSP. Similarly, space-qualified GPP follows the same QML path as FPGA and DSP, and the current on-flight rad-hard GPPs based on the following architecture are [20].

- RISC PowerPC: RAD750, RAD5500.
- RISC MIPS: RH-32, Mongoose-V, KOMDIV-32.
- Motorola 68,000 Series: Coldfire M5208
- ARM Microcontroller: Vorago VA10820

5. Summary and conclusions

In the first section of this chapter, an overview of the satellite bus and payload subsystems are presented for command and data handling subsystem (C&DHS); communications subsystem (CS); electrical power subsystem (EPS); propulsion subsystem (PS); thermal control subsystem (TCS); attitude control subsystem (ACS) also known as guidance, navigation and control (GNC) subsystem; and structures and mechanics subsystem (S&MS). A significant portion is spent on describing the C&DHS and CS with much details on how they are related to other satellite subsystems for continuous operation.

There are distinctive functional separations between the satellite bus and payload that are discussed at a high level with some examples given; however, there are currently no existing standard on their interfaces due to legacy satellite design and development. Examples were given for mission-specific sensing and communications payloads, showing that pretty much all mission payloads are very customized in design in legacy systems.

The second section of this chapter covers software defined radio (SDR) as a new technology with an overview and how SDR is being applied to satellite design and development in both space and ground segments. There has been a NASA standard for SDR that has been used for traditional and large satellites and shown to have some advantages over non-SDR approach.

However, recent rapid developments of Small Satellites (SmallSats), which CubeSat is a subset of, have resulted in an explosion of SDR applications to build

[20] For QML microcircuits, the manufacturer is required to develop a program plan that meets or exceeds the performance detailed in these appendices of MIL-PRF-38535 J standard.

[21] RHA is quantified in terms of the radiation level in Total Ionizing Dose or TID.

Pathfinder missions that can lead to successful follow-on projects. There remains to be a standard to be defined for SDR for this CubeSat application. Regardless, SDR is providing a path forward to a common framework that may enable a more generic building block for a future concept called Software Defined Satellite that will change missions based on a software upload.

Since SDR is becoming an important part of a satellite, radiation hardening of the relevant SDR components is described in some detail. The area is evolving slowly despite fast changing technology due to the additional design and manufacturing steps taken to ensure minimum effects of radiation on microelectronics. The selection of the appropriate rad-hard FPGA, DSP, and GPP components should be an important factor in design trade-offs when SDR is being considered for future missions.

Acknowledgements

The first author, Dr. Hung H. Nguyen, would like to express bountiful appreciation for his wife, Thuy Le Nguyen, for her constant support during this effort.

Author details

Hung H. Nguyen[1*] and Peter S. Nguyen[2]

1 The Aerospace Corporation, Chantilly, Virginia, USA

2 The Raytheon Corporation, Sunnyvale, California, USA

*Address all correspondence to: hung.h.nguyen@aero.org

References

[1] Pelton J, Madry S, Camacho-Lara S, editors. Handbook of Satellite Applications. Vol. 1. Springer; 2013. DOI: 10.1007/978-1-4419-7671-0

[2] Griffin M, French J. Space Vehicle Design, Second Edition, AIAA Education Series; 2004. Available from: https://arc.aiaa.org/doi/book/10.2514/4.862403

[3] Eger G III. Orion's Command and Data Handling Architecture, AIAA Space 2008 Conference & Exposition; 2008

[4] Yra P, Genna M, McMahon S, Kerns K, Tiede R, Laird M, et al. Next-Generation Spacecraft Command & Data Handling System Based on the RAD750 Processor, 28th AIAA ICSSC-2010; 2010

[5] Kobayashi M et al. NASA's high-rate Ka-band downlink system for the NISAR mission. Acta Astronautica. 2019;**159**:358-361. DOI: 10.1016/j.actaastro.2019.03.069

[6] Grayver E. Implementing Software Defined Radio. Springer; 2013. DOI: 10.1007/978-1-4419-9332-8_12

[7] Reed J. Software Radio: A Modern Approach to Radio Engineering. Prentice Hall; 2002. DOI: 10.1109/MS.2003.1207473

[8] Akeela R, Dezfouli B. Software defined radios: Architecture, state-of-the-art, challenges. Computer Communications. September 2018;**128**. DOI: 10.1016/j.comcom.2018.07.012

[9] Space Telecommunications Radio System (STRS) Architecture Standard, NASA Technical Standard NASA-STD-4009A. Approved: 2018-03-14 Superseding NASA-STD-4009 (Baseline)

[10] Quintana-Diaz G, Birkeland R. Software-Defined-Radios in Satellite Communications, ESA 4S Symposium, Sorrento, Italy, May 2018

[11] Alvarez J, Rice M, Samson R, Koets A. Increasing the Capability of CubeSat-based Software-Defined Radio Applications, 2016 IEEE Aerospace Conference, 5-12 March 2016

[12] Maheshwarappa M, Bowyer M, Bridges C. Software Defined Radio (SDR) Architecture to Support Multi-Satellite Communications, 2015 IEEE Aerospace Conference, 7-14 March 2015

[13] Asundi S, Fitz-Coy N. Design of Command, Data and Telemetry Handling System for a Distributed Computing Architecture CubeSat, IEEEAC Paper #2181, 2013

[14] Joul S, Lightsey G, Horton S, Anandayuvaraj G. A reusable command and data handling system for university cubesat missions. In: IEEE Aerospace Conference; 2014. DOI: 10.1109/AERO.2014.683636

[15] Stott D, Burek R, Eiseneich P, Kroutil J, Schartz P, Sweitzer G. The MSX command and data handling system. John Hopkins APL Technical Digest. 1996;**17**(2):143-152

[16] Hentschel T, Henker M, Fettweis G. The Digital Front-End of Software Terminals. Dredsen University of Technology; 1999. DOI: 10.1109/98.788214

[17] Sarda K, Roth N, Zee R, CaJacob D, Orr N. Making the Invisible Visible: Precision RF-Emitter Geolocation from Space by the HawkEye 360 Pathfinder Mission, 32nd Annual AIAA/USU Conference on Small Satellites, August 2018

[18] MIL-PRF-38535J. Performance Specification: Integrated Circuits (Microcircuits) Manufacturing, General Specification for (28 December 2010), US DoD

[19] Roosta R. A Comparison of Radiation-Hard and Radiation-Tolerant FPGAs for Space Applications, Jet Propulsion Laboratory, California Institute of Technology, NASA Electronic Parts and Packaging Program. 30 December 2004

[20] Radiation hardening [Internet]. 2020. Available from: https://en.wikipedia.org/wiki/Radiation_hardening [Accessed: 11 May 2020]

Chapter 2

Overview of Existing and Future Advanced Satellite Systems

John Nguyen

Abstract

This chapter presents an overview of legacy, existing, and future advanced satellite systems for future wireless communications. The overview uses top-down approach, starting with a comparison between a typical commercial regular satellite system and a high-throughput satellite (HTS) system, following by a discussion on commonly used satellite network topologies. A discussion on the design of satellite payload architectures supporting both typical regular satellite and HTS with associated network topologies will be presented. Four satellite payload architectures will be discussed, including legacy analog bent-pipe satellite (ABPS); existing digital bent-pipe satellite (DBPS) and advanced digital bent-pipe satellite using digital channelizer and beamformer (AdDBPS-DCB); and future advanced regenerative on-board processing satellite (AR-OBPS) payload architectures. Additionally, various satellite system architectures using AdBP-DCBS and AR-OBPS payloads for the fifth-generation (5G) cellular phone applications will also be presented.

Keywords: high-throughput satellite, analog bent-pipe satellite, digital bent-pipe satellite, digital channelizer and beamformer, advanced regenerative on-board processing satellite, cellular phone

1. Background and introduction

Recently, the space industry has pointed out that in the past 5 years, the commercial market has been driving the advancement of satellite technology. Lockheed Martin is building commercial satellites (e.g., Hellas-sat series) with advanced on-board processing capabilities for the Saudi Arabian [1]. Hellas satellites probably will be the first commercial HTS with a very advanced digital processor on-board. The focus of this chapter will be on commercial satellite systems for communication applications, and a comparison study between commercial HTS and typical satellites systems conducted by Inmarsat will be provided [2].

For communication applications, commercial satellite systems have been categorized as mobile satellite services (MSSs), fixed satellite services (FSSs), broadcast satellite services (BSSs), and high-throughput satellite (HTS) services. Depending on the services, satellite payload architecture will be designed to meet the specified requirements for that service. Basically, satellite payload architecture can be classified into four categories: (1) analog bent-pipe satellite (ABPS); (2) digital bent-pipe satellite (DBPS); (3) advanced digital bent-pipe satellite using digital channelizer and beamformer (AdDBPS-DCB); and (4) advanced regenerative on-board processing satellite (AR-OBPS). This chapter provides an overview of these payload architectures and presents two satellite system architectures using AdBPS-DCBS and AR-OBPS payloads for the fifth-generation cellular phone (5G) applications.

The chapter is organized as follows: Section 2 provides a comparison between commercial HTS and typical satellite systems; Section 3 discusses the typical satellite network topologies; Section 4 presents an overview of legacy ADPS transponder, existing DBPS transponder, AdBPS-DCBS transponder, and AR-OBPS satellite system; Section 5 discusses the use of AdBPS-DCBS transponder and AR-OBPS payloads for the fifth-generation cellular phone (5G) applications; and Section 6 concludes the chapter with a summary and brief discussion of way forward.

2. Typical commercial satellites and HTS comparison

Typical and regular commercial satellites are operating in C-band, Ku-band, and Ka-band with downlink frequencies approximately at 4, 12, and 40 GHz, respectively. For C-band, Ku-band, and Ka-band, the spectrum bandwidths available by geostationary orbital position are 500 MHz, 500 MHz, and 3.5 GHz, respectively. Typical antenna types for these regular commercial satellites are pointed antenna type with a single beam. Typical diameters for these pointed antennas are (a) greater than 1.8 m for C-band; (b) 0.9–1.2 m for Ku-band; and (c) 0.6–1.2 m for Ka-band satellite. **Figure 1(a)** illustrates a typical regular commercial satellite.

Typical HTSs are usually also operating in Ku-band and Ka-band with the same downlink frequencies as the regular satellites except that they employ multiple pointed beam as oppose to a single-pointed beam. **Figure 1(b)** describes a multiple beam HTS system. The salient feature of multiple beams is the frequency reuse. The frequency reuse is defined as the number of times a satellite can reuse the same spectrum and frequencies. However, high frequency reuse factor can cause potential cochannel interference or an increase in carrier-to-interference power ratio (CIR or C/I). IMMARSAT has reported that a reuse factor of 5–30 is possible with multiple spot beams employed by commercial HTS. Depending on the number of beams implemented on-board of the satellite, the cost for HTS can be twice of the cost for a regular satellite. But, the cost per bit for HTS is much lower than the regular satellite. HTS is a preferred option for point-to-point services, for example, beyond line-of-sight (BLOS) cellular phone services. **Table 1** provides a summary of the comparison of HTS and regular commercial satellites [2].

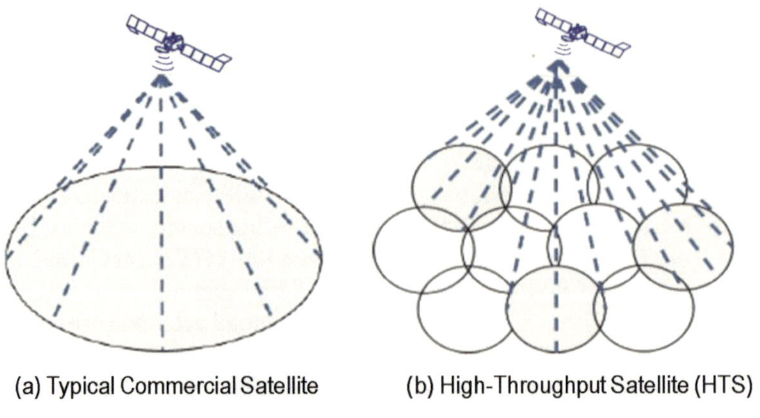

Figure 1.
Typical commercial satellites and HTS configurations.

Comparison factor	Typical regular commercial satellite	Typical high-throughput satellite (HTS)	Remark
Operational frequency band	C-band, Ku-band, Ka-band	Ku-band, Ka-band	It should be noted that for data presented here, all satellites and supply are not equal; various technical, regulatory, and commercial parameters come into play when comparing the two-type satellites. Data collected from IMMARSAT. Source: see [2]
Throughput capability (Gbps)	~1–10	~5–300+ (with frequency reuse in multiple spot beam)	
Typical cost including launch (USD)	~200–300	~300–500 (cost can be twice of regular satellite)	
Advantages	Wide coverage; preferred solution for point-to-multipoint communication	Higher bandwidth/ lower cost per bit; preferred option for point-to-point services	
Disadvantages	Limited supply available; lower spectrum efficiency for an equivalent frequency	Higher upfront costs; difficult to find enough customers to fill each of the beams	

Table 1.
Comparison of typical commercial satellites and HTS.

3. Typical commercial satellite network topologies

This section describes the most commonly used satellite network topologies, namely "Star" satellite network (Section 3.1) and "Mesh" satellite network (Section 3.2).

3.1 Typical "star" satellite network

A typical commercial satellite network topology consists of an uplink from a central anchor station (aka satellite Gateway or satellite Hub) to a satellite and a downlink from the satellite to users. Users can be mobile or fixed users. Mobile users can be located in an airplane, a boat, or a car. Fixed users can be located in a building or a cellular base station. The "star" satellite network is derived from a spoke-hub distribution paradigm in computer networks, where one central hub serves as a conduit to transmit messages among network users [3]. Thus, for star satellite networks, all communications will be passed through a satellite gateway. As shown in **Figure 2**, if Mobile User 1 wants to talk to Mobile User 2, Mobile User 1 needs to send its messages to the satellite gateway (yellow lines), and satellite gateway relays that messages to Mobile User 2 (red lines).

3.2 Typical "mesh" satellite network

The "mesh" satellite network topology is derived from a local network topology, where the network nodes are corrected to each other directly, dynamically, and nonhierarchically to as many other nodes as possible [4]. In this network topology, the network nodes can cooperate with one another to route data from one user to another user efficiently. Hence, for mesh satellite network, Mobile User 1 can talk to fixed user directly without going through the satellite gateway (solid lines), and Mobile User 2 can also talk to the fixed user directly (dash lines).

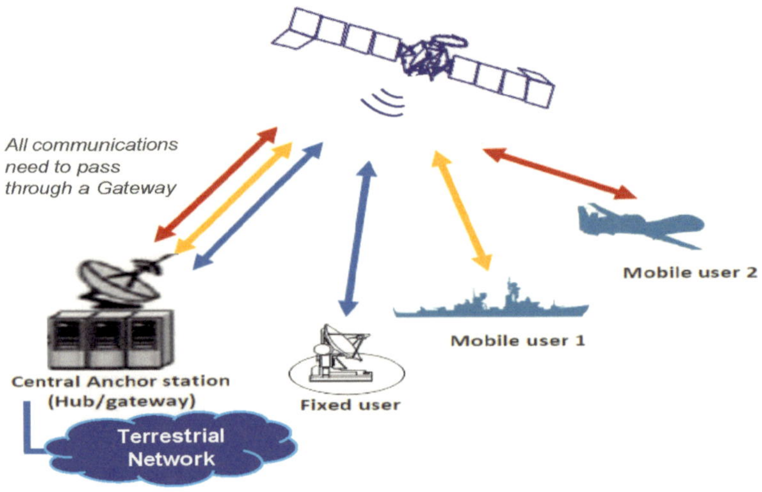

Figure 2.
Typical "star" satellite network.

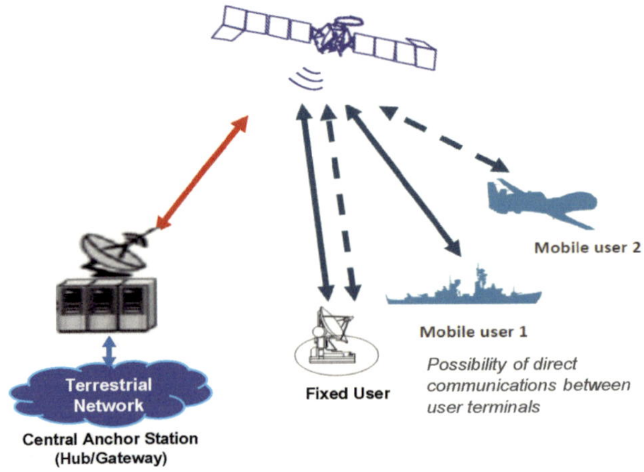

Figure 3.
Typical "mesh" satellite network.

Any one of the user within the network can send the messages to a terrestrial network through the red lines representing uplink and downlink between the satellite gateway and the satellite (**Figure 3**).

Star satellite network topology does not require advanced satellite payload processing on-board and multiple beam, but mesh satellite network requires advanced on-board processing and multiple beam allowing one user to communicate to another user automatically and effectively. Section 4 discusses various satellite payload architectures used in regular satellite and HTS for star and mesh satellite network applications.

4. Legacy, existing, and advanced satellite payload architectures

This section presents an overview of legacy, existing, and advanced satellite payload architectures. Section 4.1 presents legacy ABPS payload architecture, Section 4.2 provides a description of a typical existing DBPS payload architecture,

Section 4.3 discusses AdDBPS-DCB payload architecture, and Section 4.4 provides an overview of AR-OBPS payload architecture.

4.1 Legacy analog bent-pipe satellite (ABPS) payload architecture

A typical legacy ABPS payload architecture is depicted in **Figure 4**, where the payload has multiple beam antennas (MBAs) using parabolic dishes. For this architecture, the RF signal is received at the satellite payload and amplifies by a low noise amplifier (LNA) for increased received signal-to-noise power ratio (SNR). The RF signal with increased SNR is downconverted (D/C) to an intermediate frequency (IF) and processed by an IF filter to clean up the signal from adjacent interference and out-of-band noise. The clean-up signal is then (a) routed to the proper downlink port by an IF analog switching circuit and upconverted (U/C) to RF, (b) combined by a multiplexer (MUX), and (c) amplified by a high-power amplifier (HPA) for downlink transmission.

As illustrated in **Figure 5**, there are two options for the D/C, namely Option 1 (see **Figure 5(a)**) is a double downconverter using two local oscilators (LOs) to downconvert RF signal to IF signal with stable and low phase noise, and Option 2 (see **Figure 5(b)**) is single downconverter using a LO downconverting RF signal directlty to an IF signal. Option 1 is being used in many legacy, existing, and advanced satellite payloads. Option 2 is mostly used in advanced satellite payloads.

Figure 5(c) shows commercial-of-the-shelf (COTS) phase noise characteristics for typical LOs operating at X-band, Ku-band, and Ka-band. X-band, Ku-band, and Ka-band illustrated in this figure correspond to 7–11.2, 12–18, and 26.5–40 GHz, respectively. The main advantages of Option 2 using single downconversion are its low cost, small size, and low power consumption (also known as small SWAP-C). This option uses the smallest number of external components as compared to Option 1 using double downconversion, which is also known as super heterodyne receiver [5]. However, Option 2 suffers amplitude and phase imbalances caused by imperfect references associated with I-Q components, direct current (DC) signal due to self-mixing, and flicker noise.[1] Option 1 does not suffer from these problems and offers excellent selectivity and sensitivity, that is, better rejection of adjacent interferences. Option 1's disadvantages are the integration complexity and high SWAP-C.

In satellite electronic communications, MUX is a multiplexer, which is a device that selects several (multiple) analog (or digital) input signals and outputs a single signal. **Figure 6(a)** describes a functional MUX (aka multiplexer) circuit. On the contrary, **Figure 6(b)** depicts a DEMUX (aka demultiplexer), which is an electronic device that sends a single input signal to multiple signal outputs.

4.2 Existing digital bent-pipe satellite (DBPS) payload architecture

Figure 7 presents an existing DBPS payload architecture using on-board digital channelizer. Similar to analog payload, there are two options for the RF-to-IF downconversion process. Double-downconversion process is typically used for digital bent-pipe payload architecture.

Figure 8 depicts typical RF-to-IF (or baseband) downconversion and digitization and sampling processes for a commercial DBPS payload architecture. The RF-to-IF process shown in this figure uses Option 1, double downconversion, and the digitization and sampling process employing bandpass sampling with

[1] Flicker noise is a type of electronic noise with a 1/frequency power spectral density.

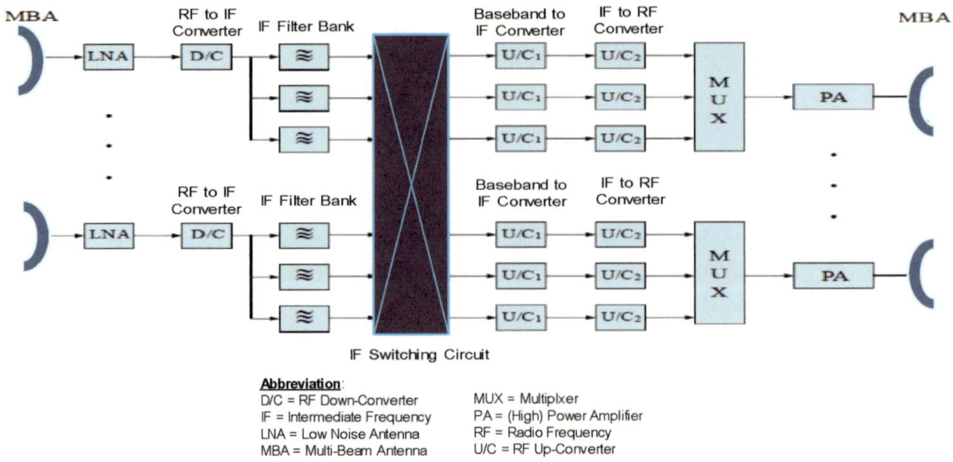

Figure 4.
Legacy ABPS payload architecture.

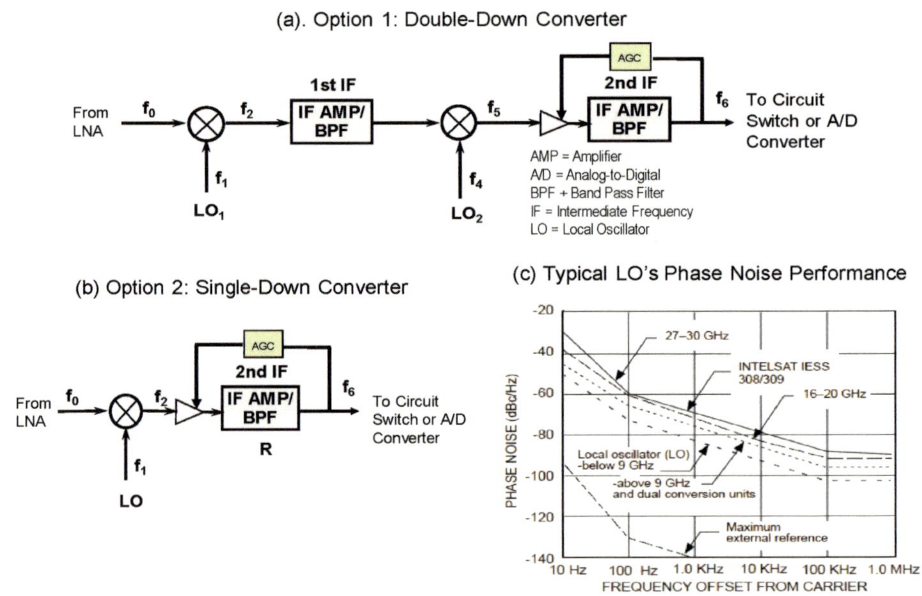

Figure 5.
Options for RF downconversion and associated LO's phase noise.

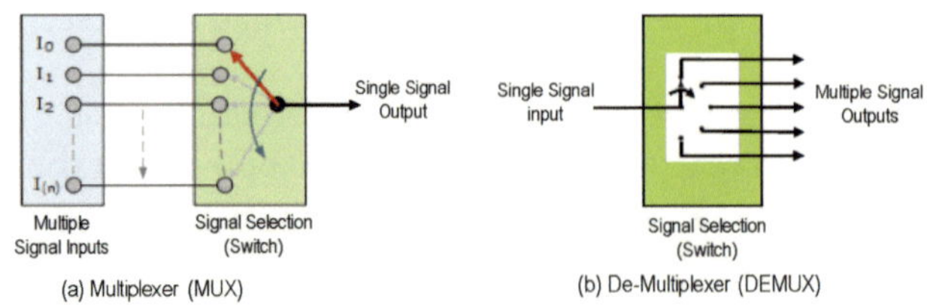

Figure 6.
Functional block diagrams of MUX and DEMUX.

digital quadrature technology [6]. The RF bandwidth (BW) associated with the RF bandpass filter (BPF) is selected to match with an over channel bandwidth (e.g., a maximum of 500 MHz for Ku-band). The automated gain control (AGC)

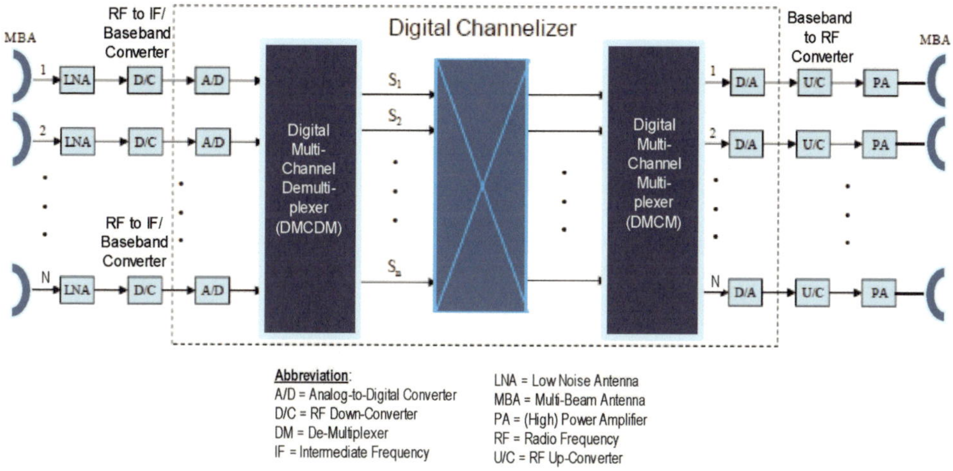

Figure 7.
Existing DBPS payload architecture.

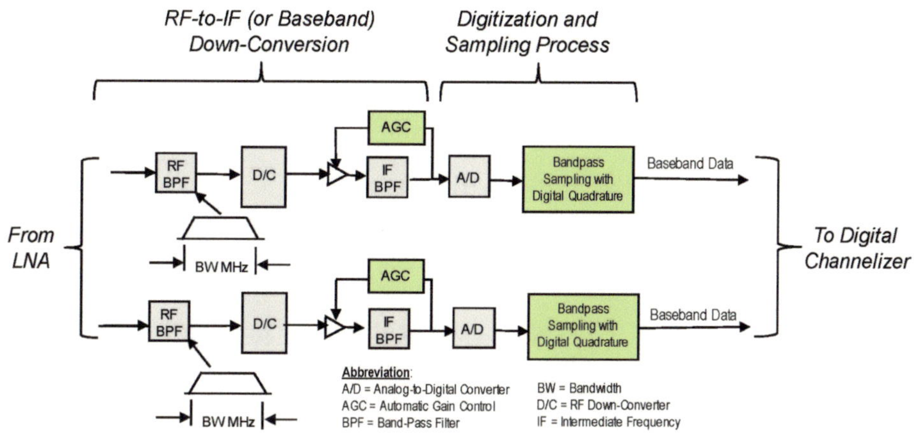

Figure 8.
Typical R/F downconversion and digitization processing approach.

is designed to maintain a constant power over the specified channel bandwidth. There are several advantages associated with bandpass sampling with digital quadrature techniques, including (a) no phase and amplitude imbalances; (b) digital finite impulse response (FIR) filters are flexible and computational complexity with linear phase introducing a constant group delay; (c) only one A/D converter is required (less weight and power); and (d) when the sampling period is set at one-quarter of the carrier frequency, the reference in-phase and quadrature components reduce to an alternating sequence between I-channel and Q-channel [6].

As shown in **Figure 9**, the key design issue associated with the digitization and sampling processing is the selection of required number of bits of the analog-to-digital (A/D) conversion to (1) achieve optimum loading factor (LF) and (2) minimize the quantization noise. The LF is defined as the root mean square (RMS) of the total input signal voltage-to-A/D converter saturation voltage ratio. The total input signal voltage includes desired signal voltage (S) plus noise voltage (N) plus interference voltage (I). **Figure 10** illustrates an optimum LF as a function of number of bit of a typical A/D converter. As an example, for 4-bit,

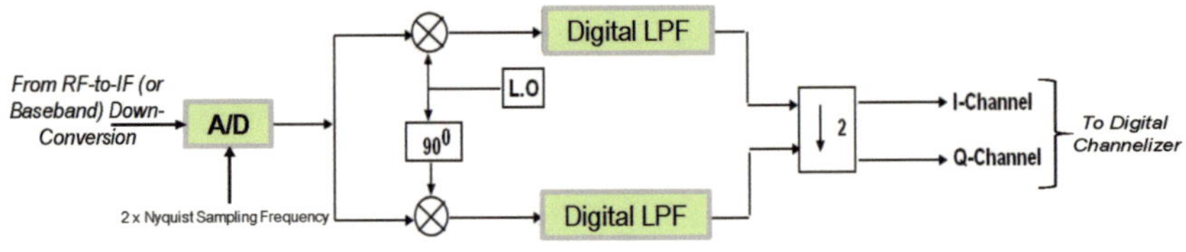

Figure 9.
Existing digitization and sampling processing using bandpass sampling with digital quadrature technique.

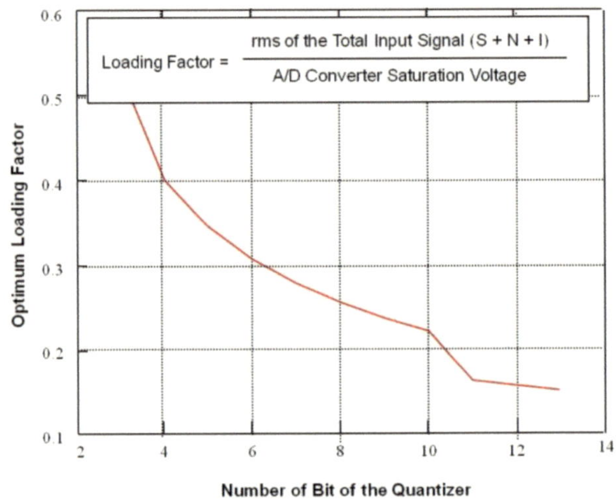

Figure 10.
Optimum LF as a function of number of bit of A/D converter.

the optimum LF is about 0.4. In conjunction with LF, the number of bit should be selected to maximize the signal-to-quantization noise ratio (SQNR) using the following relationship:[2]

$$SQNR \approx 1.761 + 6.02.N\, dB \tag{1}$$

As an example, when N = 4 bits, signal-to-quantization noise ratio is about 25.84 dB.

The key feature of DBPS payloads is the flexibility of the digital channelizer. Current digital technologies allow for the implementation of robust and reconfigurable digital channelizer adapting to require the number of users and associated users' data rates. A typical flexible digital channelizer using polyphase/discrete Fourier transform (DFT) technology is shown in **Figure 11**.

As shown in **Figure 11**, the heart of a typical digital channelizer is a polyphase-filter network (or simply a polyphase network) and a DFT processor. A typical polyphase network with a DFT processor is described in **Figure 12**. The polyphase network consists of a set of N_C digital filters with transfer function H_0, H_1..., H_{Nc-1}, which is obtained by shifting a basic low pass complex filter function along the frequency axis [7]. As an example, for a typical 500 MHz channel bandwidth, assuming for a typical user data rate of 4 MHz and a guardband of 1 MHz, digital channelizer, N_C = 500/(4 + 1) = 100, that is, the number of filter is 100, and each has a total of 5 MHz bandwidth. A change in sampling frequency by a factor of N_C can

[2] Quantization (signal processing). Available from: https://en.wikipedia.org/wiki/Quantization_(signal_processing).

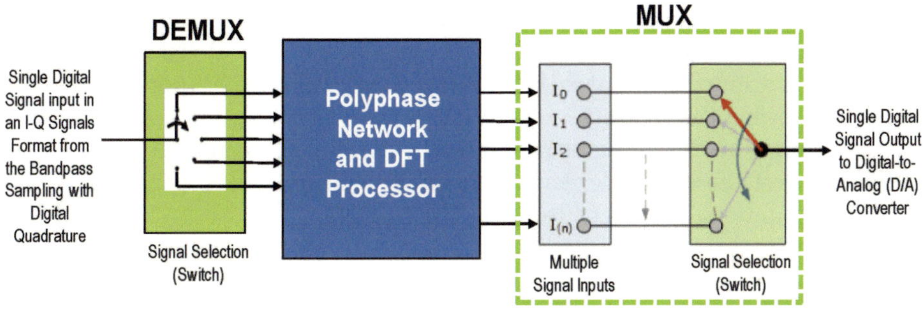

Figure 11.
Typical digital channelizer using polyphase/DFT technology.

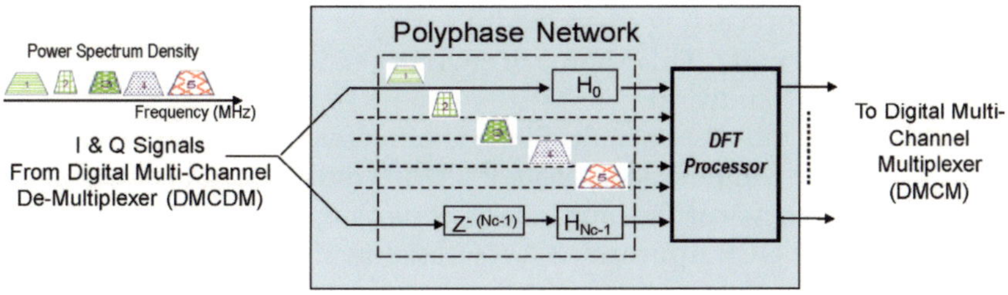

Figure 12.
Typical Polyphase/DFT Technology.

be introduced, thus allowing the circuit in different paths of the polyphase network to operate at lower frequency than the original sampling frequency. A practical implementation of a high-throughput low-latency polyphase channelizer can be found in [8, 9].

Figure 12 shows an example of five input signals, namely S_1, S_2, S_3, S_4, and S_5, and the channelizer will select signal interest by filtering out the other signals. As an example, the signal line with the filter transfer function of H_0 filters out S_2, S_3, S_4, and S_5 and sends S_1 as an output signal.

4.3 Advanced digital bent-pipe satellite using digital channelizer and beamformer (AdDBPS-DCB) payload architecture

For a typical commercial HTS system architecture, it usually requires on-board multiple beam phase array (PA) antenna with associated adaptive digital beam-former network (DBF) for spot beamforming and frequency reusing of the spot beams when the beams are not located near each other. **Figure 13** describes a typical AdDBPS-DCB payload architecture, where the digital channelizer is combined with a DBF to make a "digital channelizer and beamformer" (DCB) [10–12]. For this payload architecture, the key feature that differentiates this architecture with the ones discussed above is the combined digital channelizer using polyphase network/ DFT processor and DBF (PolyN/DFT-DBF).

As pointed out in [10–12], DCB architecture shown in **Figure 13** can be designed to (1) form individual beams for each active receive and transmit communication channels; (2) adaptively generate channel beam steering weights to dynamically vary the bandwidth, location, and shape of each beam based on traffic demands and the locations of other, potentially interfering beams avoiding adjacent channel interference; (3) use digital beamforming weight calibration to compensate for the temporal and thermal phase and amplitude response variations inherent in analog multibeam phased array antennas; and (4) adjust the gain of individual receive-and-transmit channel beams automatically to compensate for propagation path and analog payload response variations. In general, there are two possible DCB implementation approaches, namely DCB Approach 1 and DCB Approach 2 [13]. **Figure 14** describes the DCB Approach 1 for processing the uplink signals, where the uplink signals are individually processed by the digital channelizer (i.e., PolyN/DFT processing) and DBF independently and separately. DCB Approach 1 requires a larger computational load because each DBF processes all the user link bandwidth (e.g., S_1, S_2, S_3, S_4, and S_5 in **Figure 12**) at all times to form multiple beams.

DCB Approach 2 is shown in **Figure 15**, where DCB utilizes an unified processing approach with each DBF processes only the bandwidth corresponding to a beam (S_1 in **Figure 12**) at normal times. During anomaly operation condition (e.g., natural disaster event), when the bandwidth has to be reassigned to specific areas, the arithmetic load on DBF can be reduced by implementing multiple DBFs, with each capable of processing a bandwidth narrower than that assigned to a beam (i.e., smaller channel unit). This approach enables a reduction in wasteful arithmetic resource usage on bandwidth.

If one defines the number of multipliers, D implemented in each Tx/Rx DBF as C/fop, where C is the computational load of a DBF (multiplications/sec), and fop is the operation frequency of the multiplier. Let us compare D calculations between DCB Approach 1 and DCB Approach 2. Let us assume the following parameters: n is the number of array elements, m is the number of beams, an userlink processing bandwidth of 28 MHz, 5 frequency repetitions of the userlink, and an operating

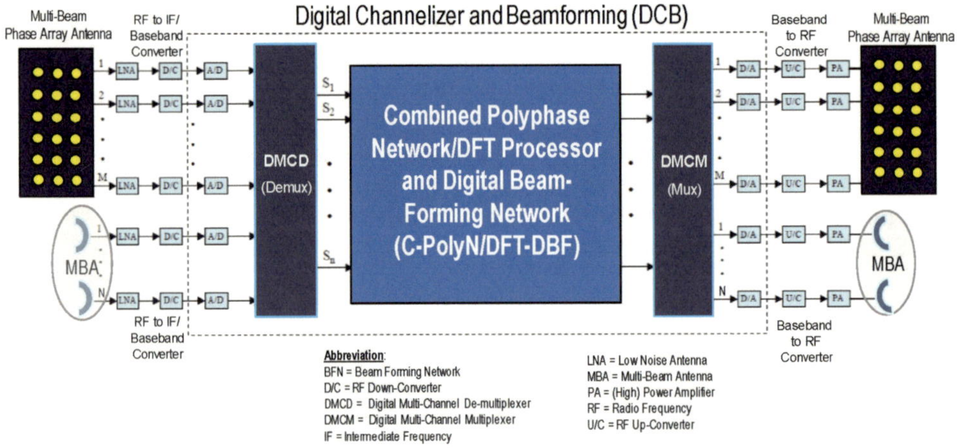

Figure 13.
AdDBPS-DCB payload architecture.

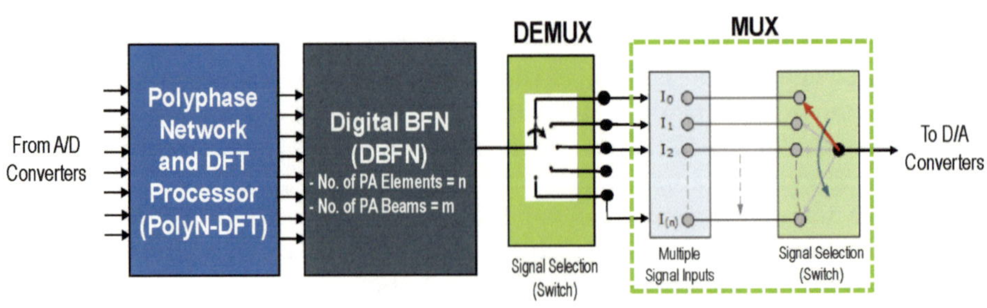

Figure 14.
DCB Approach 1: PolyN/DFT and DBFN individual processing.

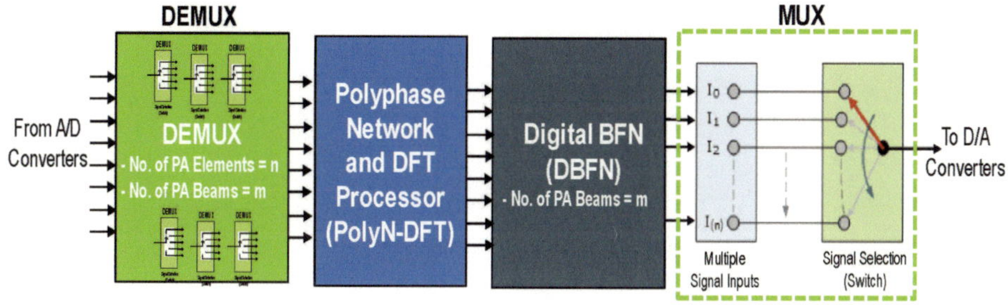

Figure 15.
DCB Approach 2: Unified and combined PolyN/DFT and DBFN individual processing.

frequency of multiplier of 256 MHz. Using these values, D for the DBF/channelizer of the DCB Approach 1 configuration becomes [13]:

$$\frac{\left(n \times 4 \times m \times 28 * 106\left[\text{multiplications}/\text{s}\right]\right)}{\left(256 * 106\left[\text{multiplications}/\text{s}\right]\right)} \tag{2}$$

and that for DCB Approach 2 configuration becomes [13]:

$$\frac{\left(n \times 4 \times m \times 28 * 106\ /5\left[\text{multiplications}/\text{s}\right]\right)}{\left(256 * 106\left[\frac{\text{multiplications}}{\text{s}}\right]\right) \times 2} \tag{3}$$

The latter calculation assumes an ideal case in which DBF network (DFBN) processing is performed on a channel-by-channel basis. The complexity of DCB Approach 2 configuration is 10 times less complex than DCB Approach 1.

As pointed out in [12], the DBFN when coupled with a digital channelizer (aka DCB) offered more capabilities with many advantages. Nguyen et al. [14] developed a computer simulation model of a typical DBFN in MATLAB and presented simulation results for X, Ku, and Ka BFNs using 60-element, 104-element, and 149-element, respectively. **Figure 16** is an extracted Ka-band BFN result showing the achievable antenna gain of 45.5 dB at 3-dB beamwidth of 0.9°. For practical applications, the DBFN will shape the beam size depending on the coverage area and desired number of beams. Nguyen et al. [14] pointed out that for 2.5° coverage area and the desired number of beams of 7, the minimum 3-dB beamwidth of 1.1° is required. Nguyen et al. [14] also pointed out that DCB can provide a significant increase in frequency reuse, where the frequency reuse is defined as the number of times a satellite can reuse the same spectrum and frequencies. High frequency reuse factor can cause potential cochannel interference (CCI) that results in a decrease in carrier-to-interference power ratio [aka (C/I) CCI]. As pointed out in [14], for dynamic allocation using real-time allocation of beams so that the coverage radius of a cell is equal to the satellite pointing error, assuming satellite pointing error of 0.02 degree pointing error, the (C/I)CCI is about 25 dB for frequency reuse factor 40 [14].

4.4 Future advanced regenerative on-board processing satellite (AR-OBPS) payload architecture

Figure 17 depicts a potential future AR-OBPS payload architecture [10]. The payload includes (1) a typically set of digitized analog multiple beam antenna (MBA) input signals, digitally frequency division demultiplex each input signal to produce single carrier per channel (SCPC) signal data and demodulate and decode

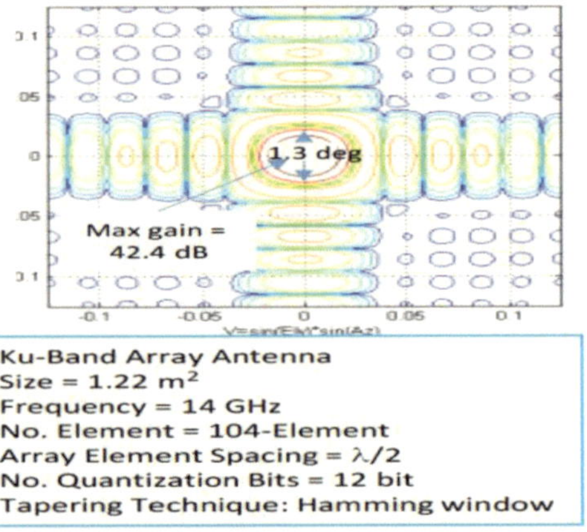

Figure 16.
Antenna beamwidth and gain of a notional Ka-band DBFN with 12-bit quantization [14].

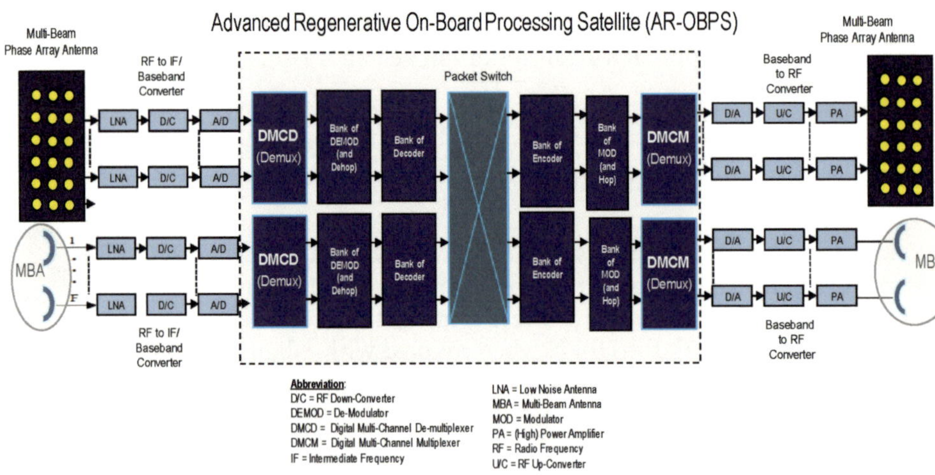

Figure 17.
AR-OBPS payload architecture.

individual traffic channels to recover the original information bits transmitted on the uplink; (2) a set of digitized analog multibeam phase array antenna (MB-PAa) input signals, digitally frequency division demultiplex each input signal to produce SCPC signal data and demodulate and decode individual traffic channels to recover the original information bits transmitted on the uplink; and (3) fast packet switches are typically employed at the AR-OBPS payload's core to realize statistical multiplexing gains by efficiently packing and moving data through the switch and onto the downlink in bursty uplink transmission applications. Moreover, the digital bandwidth (in Hz) through the AR-OBPS switch is at least 25 times less[3] than that supported by an equivalent (pre-demodulation) digital baseband switch at the center of a DC- or DCB-based system. AR-OBPS payload can also support digital beamforming, following the frequency division demultiplexing operation, if a phased array is employed in place of the analog MBA. On the secondary (output) side of the switch, each user's binary information is channel encoded and modulated onto a carrier. The modulated carrier data thus produced are multiplexed,

[3] Assumes 1 bps/Hz modulation efficiency, 10 bit signal data quantization, and 2.5× practical Nyquist sampling rate.

digital-to-analog converted, and passed through an analog reconstruction filter to generate output signals for the transmit portion of the communication payload. The channel codes and modulations employed on the uplink (input) communication channels clearly do not need to be the same as the channel codes and modulations used on the transmitted downlink channels. Hence, an AR-OBPS payload can serve as a "translator" facilitating single-hop communications between terminals employing different link protocols. However, if either the digital multichannel demultiplexer (DMCD), demodulator, decoder, or digital multichannel multiplexer (DMCM) encoder modulator, multiplexer (MCEM2) functions are implemented in ASICs to minimize size-weight-and-power (SWaP), then the AR-OBPS system becomes somewhat inflexible, unable to support either uplink or downlink terminals, respectively, using communication protocols differing from those for which the AR-OBPS was specifically designed. For this reason, AR-OBPS systems are typically employed in support of "private networks" in which the communication satellite service provider only accommodates terminals designed to work on the provider's network. Iridium and Spaceway are two examples of commercial AR-OBPS-based communication satellite systems.

5. Satellite system architectures for 5G cellular phone applications

Sections 5.1 and 5.2 present a notional satellite system architecture using AdBPS-DCBS satellite payload and AR-OBPS satellite system architecture for 5G cellular phone applications, respectively.

5.1 AdBPS-DCBS satellite system architecture for 5G applications

AdBPS-DCBS satellite payload can be used to support 5G users. There are potentially two satellite system architecture options for using AdBPS-DCBS satellite payload to support 5G mobile user equipment (aka 5G-UE), namely AdBPS-DCBS Option 1 and AdBPS-DCBS Option 2. For AdBPS-DCBS Option 1, the AdBPS-DCBS satellite provides communication services directly to 5G-UEs. While in AdBPS-DCBS Option 2, the satellite provides services to 5G-UEs through the 5G relay nodes (RNs). **Figure 18** illustrates the AdBPS-DCBS satellite system architecture for (a) AdBPS-DCBS Option 1 and (b) AdBPS-DCBS Option 2 [15].

Figure 18(a) shows that the AdBPS-DCBS satellite requires new radio (NR) interfaces between (1) AdBPS-DCBS satellite and terrestrial gateway (GW) and (2) AdBPS-DCBS satellite and 5G-UEs. In addition, it is also required a 5G narrow-band (gNB) processing station to process the 5G signals from the next generation core (NGC) network before passing the 5G data to public data network.

5.2 AR-OBPS satellite system architecture for 5G applications

Similar to AdBPS-DCBS satellite payload, AR-OBPS satellite payload can also be used to support 5G users. There are also two satellite system architecture options for using AR-OBPS payload to support 5G mobile user equipment, namely AR-OBPS Option 1 and AR-OBPS Option 2. For AR-OBPS Option 1, the AR-OBPS satellite provides communication services directly to 5G-UEs. For AR-OBPS Option 2, the satellite provides services to 5G-UEs through the 5G RNs. **Figure 19** describes these two AR-OBPS architecture options, namely (a) for AR-OBPS Option 1 and (b) for AR-OBPS Option 2. For these two system architecture options, the gNB processing is now incorporated into the AR-OBPS satellite payload and no longer required for the ground system. The GW now can pass the 5G data directly to the NGC. The

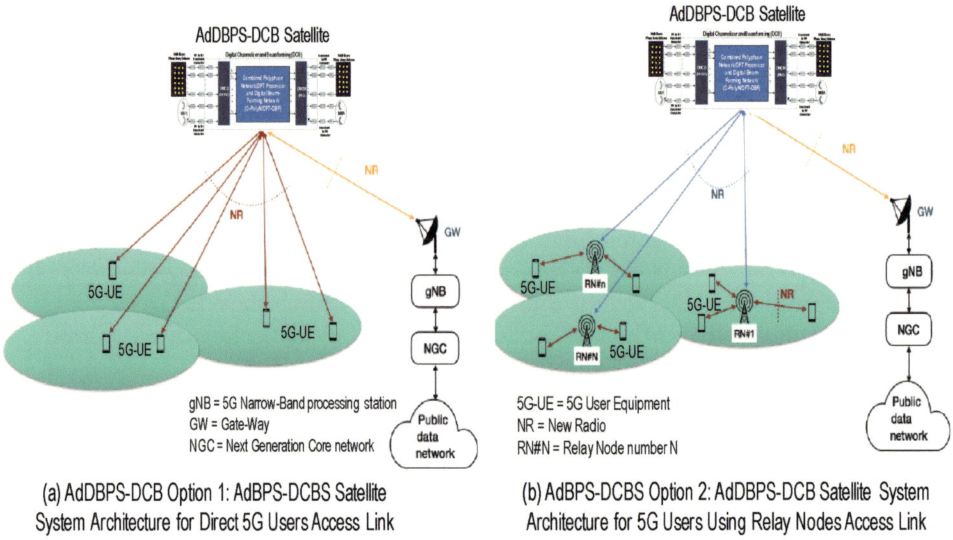

Figure 18.
AdDBPS-DCB satellite system architectures for supporting 5G users.

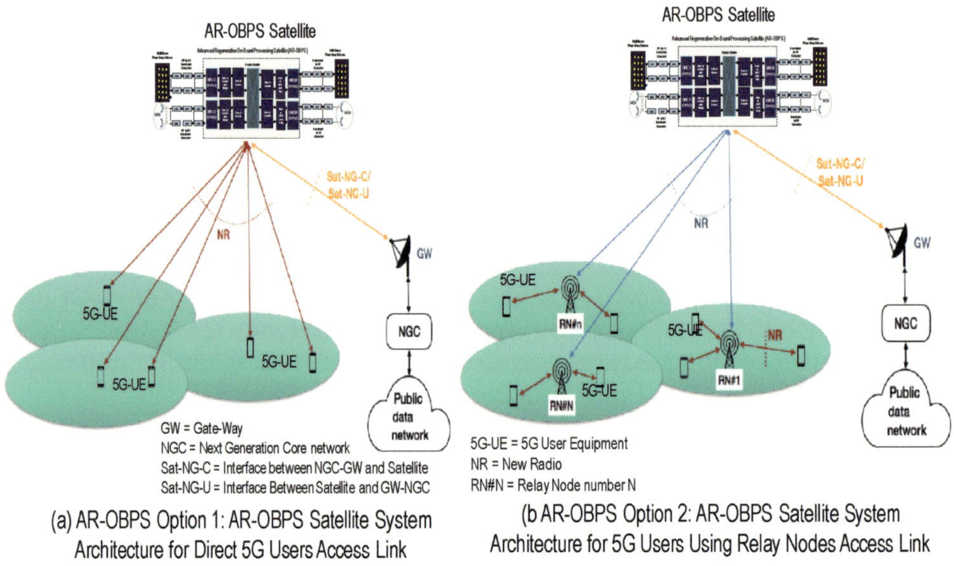

Figure 19.
AR-OBPS satellite system architectures for supporting 5G users.

decoding-demodulation and encoding-modulation processing on-board of the satellite will be designed to align with the 5G waveform specifications, including 5G modulation and coding schemes.

Figure 19(a) shows that the AR-OBPS satellite also requires NR interfaces between (1) AR-OBPS satellite and GW and (2) AR-OBPS satellite and 5G-UEs. Similar to AdBPS-DCBS satellite system architecture options, the NR interfaces between the AR-OBPS satellite and 5G-UEs are new. Since the gNB processing is now placed at AR-OBPS satellite payload, the NR interfaces between AR-OBPS satellite and 5G-UEs are not the same as the AdBPS-DCBS satellite and 5G-UEs. To show the differences between the two, **Figures 19(a)** and **(b)** use Sat-NG-C and Sat-NG-U to indicate the new radio interface between (1) terrestrial GW-NGC-and-AR-OBPS satellite and (2) AR-OBPS satellite-and-terrestrial GW-NGC, respectively.

6. Conclusion

This chapter uses a top-down approach for providing an overview of legacy, existing, and future advanced satellite payload architectures for future wireless communication applications. The chapter focuses on the commercial satellite technologies based on the research results presented in [1, 2]. Section 2 provides the comparison results performed by Inmarsat describing the technical characteristics and associated advantages and disadvantages between commercial HTS and typical satellite systems currently available in commercial satellite market. In Section 3, two most commonly satellite network topologies used by existing commercial satellite networks are presented, and the concept of satellite uplink and downlink associated with star satellite network and mesh satellite network is discussed. The satellite network topologies presented lead to Section 4, where four satellite payload architectures are discussed. The legacy analog ABPS payload architecture is shown to be more appropriate for star satellite network than mesh network. Existing digital DBPS and AdDBPS-DCB payload architectures are designed for supporting mesh satellite network with large number of mobile users. Future advanced digital satellite payload architecture, namely AdDBPS-DCB, is also presented in this section. With decoding-demodulating and encoding-modulating processing on-board of the satellite, AR-OBPS allows for packet switching on-board and higher quality of service (QOS) than existing DBPS and AdDBPS-DCB at the expense of higher SWAP Cost (SWAP-C). Section 4 of the chapter discusses the applications of AdBPS-DCBS and AR-OBPS payloads for supporting 5G users. Four satellite system architecture options are presented for supporting the future 5G users.

Conflict of interest

The preparation of this chapter was not funded by Gulfstream, and it was done by the author using his own time and resources; thus, it does not represent the Gulfstream's view on the results presented in this chapter.

Notes/Thanks/Other declarations

The author wishes to thank his wife, Annie Luu-Nguyen, for her immense patience and support.

Author details

John Nguyen
JohnDTN Consulting Services, Huntington Beach, California, USA

*Address all correspondence to: johndncva@gmail.com

References

[1] Magnuson S. Commercial space—Not military—Driving satellite innovation. National Defense Magazine. 2018. Available from: https://www.nationaldefensemagazine.org/articles/2018/4/17/commercial-space-not-military-driving-satellite-innovation

[2] Revillon P. Fundamentals and Dynamics of the Satellite Communications Business. Presented at the Euroconsult for Inmarsat Capital Markets Day in 2016. Available from: www.euroconsult-ec.com

[3] Star Network. Wikipedia. Available from: https://en.wikipedia.org/wiki/Star_network

[4] Mesh Network. Wikipedia. Available from: https://en.wikipedia.org/wiki/Mesh_networking

[5] Superheterodyne Receiver. Wikipedia. Available from: https://en.wikipedia.org/wiki/Superheterodyne_receiver

[6] Sadr R, Shahshahani M. On Sampling Band-Pass Signals, TDA Progress Report 42-96. Pasadena, California: NASA-Jet Propulsion Laboratory; 1988

[7] van der Veldt K, van Nieuwpoort R, Varbanescu AL, Jesshope C. A polyphase filter for GPUs and multi-core processors. In: Proceedings of the Workshop on High-Performance Computing for Astronomy; Delft, Netherlands; 18 June, 2012. pp. 33-40

[8] Kim SC, Bhattacharyya SS. Implementation of a high-throughput low-latency polyphase channelizer on GPUs. In: EURASIP Journal on Advances in Signal Processing; 2014. Available from: http://asp.eurasipjournals.com/content/2014/1/141

[9] Ejima F, Akita M, Fujimura A. Digital Channelizer for High Throughput Satellite Communications, Technical Paper. Mitsubishi Electric Advance (Technical Journal published by Mitsubishi); 2014. pp. 7-10

[10] Butash TC, Marshall JR. Leveraging digital on-board processing to increase communications satellite flexibility and effective capacity. In: 28th AIAA ICSSC. AIAA 2010-8715; Anaheim, CA; 30 August–2 September 2010

[11] Sichi SF, Ziegler HE. Beamforming architectures for advanced MSS network deployment. In: 29th AIAA ICSSC; AIAA 2011-8021; Nara, Japan; 28 November–01 December, 2011

[12] Freedman JB, Marshack DS, et al. Advantages and capabilities of a beamforming satellite with a space-based digital processor. In: 32nd AIAA International Communications Satellite Systems Conference (ICSSC); AIAA 2014-4321; San Diego, CA; 4-7 August 2014

[13] Komiyama N, Miura A, Orikasa T, Fujino Y. Development of resource allocation re-construction technology (digital beam former and digital channelizer). Journal of the National Institute of Information and Communications Technology. 2015. pp. 151-163

[14] Nguyen TM, Guillen A, Matsunaga S. Practical achievable capacity for advanced SATCOM on-the-move. In: 2016 IEEE MILCOM Conference Proceedings; 2016

[15] Guidotti A, Vanelli-Coralli A, Conti M, Andrenacci S, Chatzinotas S, Maturo N, et al. Architectures and key technical challenges for 5G systems incorporating satellites. IEEE Transactions on Vehicular Technology. 2019. pp. 2624-2639

Section 2

Satellite and Space Control

Chapter 3

Satellite Control System: Part I - Architecture and Main Components

Yuri V. Kim

Abstract

This chapter provides introductory material to satellite control system (SCS). It is based on the author's experience, who has been working in areas of SCS development, including designing, testing, operating of real SCS, as well as reviewing and overseeing various SCS projects. It briefly presents SCS generic futures and functional principles: tasks, architecture, basic components and algorithms, operational modes, simulation and testing. The chapter is divided into two parts, namely, Part I: SCS Architecture and Main Components and Part II: SCS Simulation, Control Modes, Power, Interface and Testing. This chapter focuses on Part I. Part II will be presented as a separate chapter in this book.

Keywords: satellite control, attitude and orbit, determination, estimation, sensors, actuators, coordinate systems, reference frame, state estimation and Kalman filtering, earth gravity, magnetic fields

1. Introduction

Satellite control system (SCS) is a core, essential subsystem that provides to the satellite capabilities to control its orbit and attitude with a certain performance that is required for satellite mission and proper functioning of satellite payload operation. However, the first mandatory task for SCS is assuring satellite safe functionality; providing sufficient electric power, thermal and communication conditions to be able for nominal functioning during specified life time at different sun lightening conditions (including potential eclipse periods), protecting against life critical failures proving to satellite safe attitude in Safe Hold Mode (SHM). Without SCS or satellite guidance, navigation and control (GN&C) system, any Earth-orbiting satellite could be considered just as artificial space body, demonstrating the *launcher* capability for the satellite launch. As soon as a satellite is assigned to perform a certain space mission, it has to have SCS and a kind of special device (s)-payload (s), performing scientific, commercial or military tasks that are dedicated to this mission. Today, the widespread satellite and SCS design philosophy [1–3] is based on the concept that satellite is a platform (bus or transportation vehicle) for the very important person (VIP) passenger, which is the payload, and this platform is aimed just to deliver and carry it in space. This approach has been proven as successful or, at least, satisfactory from the commercial point of view. However, the first Soviet satellite "Sputnik" and further Soviet/Russian satellites were built and launched under the different philosophy that satellite is the main "personage" performing a space mission and the payload (unlikely the ballistic rocket war head (s)) is just one of the satellite subsystems that should be integrated into the satellite board under the satellite chief designer guidance, who is responsible for the mission performance. From the author's point of view, this approach has certain advantages following from the Aerospace System Engineering, integration and distribution functions, and responsibilities between the space mission participants. In this chapter, SCS is presented from this point of view, integrating conventionally separate satellite GN and C subsystems and devices into the joint integrated system, attitude and orbit determination and control system (AODCS). The main principles and features of this system are presented in this chapter.

2. Earth-orbiting satellites and the role of the control system

The first human-made Earth-orbiting satellite (Soviet Sputnik), Simplest Satellite (SS-1), was launched on October 4, 1957. This satellite was launched following the development of the Soviet intercontinental ballistic rocket R-7 (8 K71). Nevertheless, it started a new era of space human exploration (**Figure 1**).

SS-1 technical characteristics are as follows [4, 5]:

- Mass 83.6 kg; sealed from two identical hemispheres with a diameter of 0.58 m; life time 3 months; payload, two 1 W transmitters (HF, 20.005 and VHF, 40.002 MHz) with four unidirectional deployable antennas (four 2.4–2.9 m metallic rods); electrical batteries, silver-zinc; sufficient for 2 weeks.

- Orbit: perigee 215 km, apogee 939 km, period 96.2 min, eccentricity 0.05, inclination angle 65.10 deg.

- Inside, the satellite sphere was filled by nitrogen, and the temperature was kept within 20–23 deg. C with automatic thermoregulation-ventilation system (thermometer-ventilator).

- The satellite had no attitude control and was free rotated around its center of mass in orbit, keeping initial angular speed, provided by the separation pulse after the separation from the launch rocket. However, thanks to the four rod

SS1 launch rocket General Designer

S.Korolev (R-7)

SS1 Lead Designer

M. Homiakov

Figure 1.
Soviet designers-creators of the first earth-orbiting artificial satellite SS-1.

antennas that provided unidirectional radio transmission in the two-radio bends, HF and VHF, SS-1 evidently indicated its presence in space for all people over the world. Even amateur radio operators with amateur receivers could receive famous now signals: BIP, BIP, BIP ... !! (**Figure 2**).

Figure 2.
SS-1, assembled (left). Open two semispheres (right).

Since SS-1, about 8378 satellites were launched to year 2018 [6]. Early satellite launches were extraordinary events and demonstrated tremendous achievement of the launched state, the USSR (4 Oct. 1957, SS-1), the USA (31 Jan. 1958, Explorer 1) and Canada (29 Sep. 1962, Alouette, launched by Thor-Agena, a US two-stage rocket), but with time, satellite launches became ordinary and usually pursue a certain military or civil mission.

Among the civil missions (satellites), the following types can be determined as already conventional: navigation, communication, Earth observation, scientific, geophysics and geodetic, technology demonstration and developers training. These satellites are usually equipped with a kind of payload system(s) (radio/TV transmitter/transducer, radar, telescope or different scientific instrument, etc.) to perform certain dedicated space mission(s). For example, the first Canadian Earth observation satellite RADARSAT-1 (Nov 4, 1995–May 10, 2013; **Figure 3**) was equipped with a side-looking synthetic aperture radar (SAR) on board the International Space Station (November 1998, ISS; **Figure 4**) was installed a Canadian robotic arm for its assembling and maintenance.

Figure 3.
The first Canadian earth observation satellite RADARSAT-1.

According to the satellite altitude (h), their orbits can be classified as low-altitude (LEO), 200–2000 km; medium-altitude (MEO), 5000–20,000 km; and high-altitude (HEO), h > 20,000 km; according to eccentricity as: close to circular $e < 0.01$; elliptical $0.01 < e < 0.3$; highly elliptical $0.3 < e < 0.8$.

There are satellites with special type of orbit such as polar (i = 90 deg), equatorial geostationary (GEO, i = 0 and h = 35,800 km) and Sun-synchronous provide orbital precession equal to Sun annual rate (i depends on satellite period) (**Figure 5**).

Figure 4.
International Space Station (ISS).

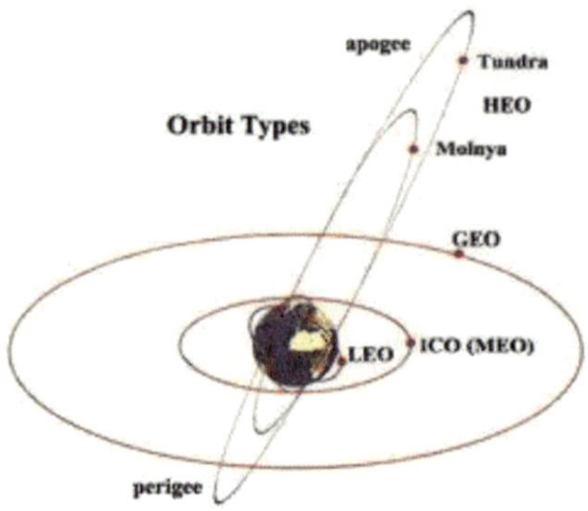

Figure 5.
Satellite orbit types ("tundra" and "Molniya" are Russian communication satellites in highly ecliptic orbits).

Miniaturized low-cost satellites are as follows: small satellites (100–500 kg), microsatellite (below 100 kg) and nanosatellite (below 10 kg).

A large diversity of satellites serving for different missions is in space now. A widespread point of vew is that all of them are transportation platforms delivering and carrying in orbit dedicated to the planned space mission payload system, like a VIP passenger. For example, it could be the postman for the postal horse carriage for many years ago. Namely, the satellite with its control system (SCS) provides to the payload all conditions required for the mission performance (orbit, attitude, power, pressure, temperature, radiation protection and communication with ground mission control center (MCC)). That is why from the mission integration point of view, the SCS can be seen as the space segment integration bases that set their development and operation process in corresponding order. In turn, SCS as satellite subsystem also can be reviled and established in satellite onboard equipment architecture, combining the group of subsystems that are dedicated to orbit and attitude determination and control tasks. It could be done rather from the System Engineering than from the commercial practice point of view and would significantly streamline satellite development order and the degree of responsibility of all the developers.

It should be mentioned that such group of aircraft equipment in aviation has been named as GN&C Avionics; hence, for space, it can be named as the *Spacetronics*, and the heritage of system development and integration wherever it is possible should be kept. Essential difference with Avionics for the Spacetronics is that it should work for specified life time in space environment (dedicated orbit) after mechanical start-up impacts (overload, vibration) connected to the launch into the orbit. The verification of this capability is usually gained in special space qualification ground tests that imitate launch impact and space environment with thermo-vacuum and radiation chambers, mechanical load and vibration stands [7, 8].

3. Satellite control system architecture and components

3.1 SCS architecture

Today, for many satellites, GN&C onboard equipment can be presented by the following subsystems, performing related functions listed below:

- Global Positioning System (GPS)—onboard satellite orbit and time determination

- Propulsion system—orbit/attitude control system

- Attitude Determination and Control System (ADCS)—satellite attitude determination and control

Integration of these subsystems can be named as *attitude and orbit determination and control system or Spacetronic system.* Typically, AODCS includes the following components:

- Onboard computer system (OBCS) or dedicated to AODCS electronic cards (plates) in Central Satellite Computer System (e.g., command and data handling computer (C&DH))

- Sensors

- Actuators

Basic AODCS architecture is presented in **Figure 6**.

OBCS, onboard computer system; TLM, telemetry data and commands; PL, payload; PS, propulsion system; RW, inertia reaction wheels; MTR, magnetic torque rods; GPS, satellite navigation Global Positioning System; MAG, 3-axis magnetometer; SS, 2-axis Sun sensor; HS, horizontal plane sensor; ST, star tracker; RS, angular rate sensor; EP, electric power; TR, temperature regulation; VP, vacuumed protection; RP, radiation protection.

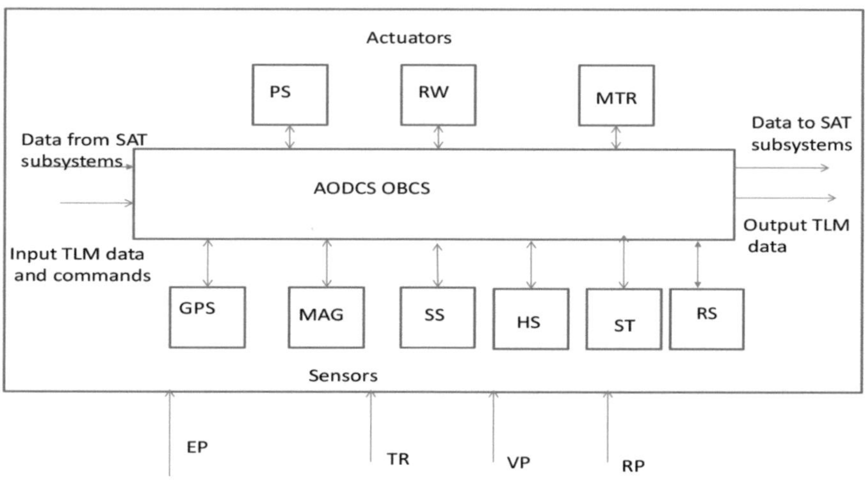

Figure 6.
Satellite AODCS system.

Depending on required reliability and life time, each component can be a single or redundant unit. Unlike airplanes, satellite is an inhabitant space vehicle that is operated from the ground. The operation is usually performed via a bidirectional telemetry radio link (TLM) in S-band (2.0–2.2 GHz). Payload data downlink radio link (unidirectional) is usually performed via X-band (7.25–7.75 GHz;). For both links, usually the same data protocol standards are applied **Figure 7**.

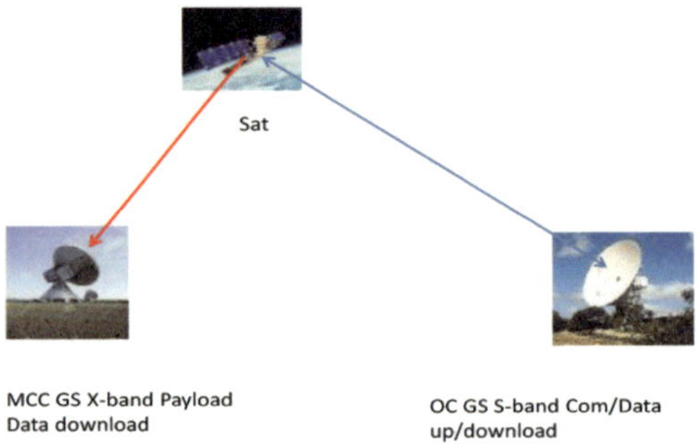

Figure 7.
Satellite communications with ground stations.

Two subsystems can be allocated in AODCS architecture, namely, orbit determination and control subsystem (ODCS) and attitude determination and control subsystem (ADCS). Practically both subsystems are dynamically uncoupled; however, orbital control requires the satellite to have a certain attitude (as well as orbital knowledge itself), and attitude control requires orbit knowledge also. Hence, orbit (its knowledge) is essentially continuously required on satellite board where it is propagated by special orbit propagator (OP). Due to orbital perturbations (residual atmospheric drag, gravity and magnetic disturbances and solar pressure), satellite orbit changes over time and OP accumulates errors; its accuracy is degraded.

Before the application of satellite onboard GPS receivers, the satellite position and velocity were periodically determined on ground by the ground tracking radio stations (GS, dish antenna), and calculated on-ground orbital parameters were periodically uploaded to satellite OBCS to correct OP, to provide available accuracy. Now with GPS satellite, orbit can be calculated onboard autonomously, and OP can propagate data only during relatively short GPS outage periods. For some applications, orbital data uploaded from the ground still can be used, at least, for fusion with GPS-based OP.

For newly developed satellites with GPS, orbit maneuvers (correction, deorbiting, collision avoidance, special formation flying and orbit servicing missions) can be executed autonomously onboard at planned time or from ground operators using orbital knowledge and TLM commands to activate satellite orbit control thrusters.

3.2 AODCS components

Below AODCS components are presented to show their generic principles that can help for the system understanding and modeling. Generic design requirements are presented in [3]. Some design examples can be found in many sources, for example, [1, 9–12].

3.2.1 Sensors

AODCS sensors are designed to measure satellite orbital and attitude position and velocity. From the most general point of view, they can be considered as the vector measuring devices (VMD). The device can measure in space a physical vector $\overline{R}_m$ that can be known (referenced) in a reference coordinate system $\overline{R}_r$. Three parameters can be measured: vector module R and two angles of its orientation Az and El (**Figure 8**).

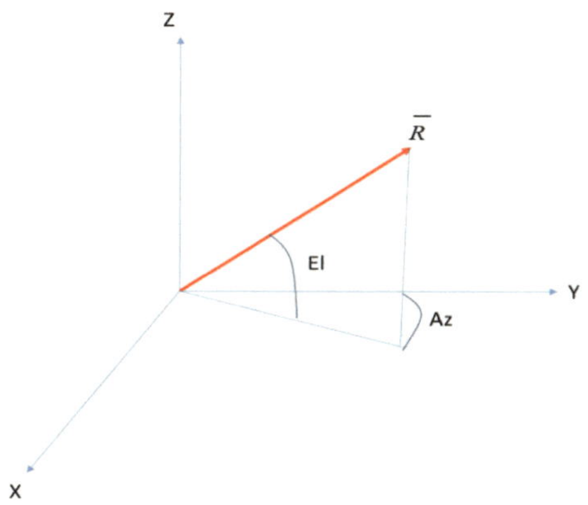

Figure 8.
Vector $\overline{R}$ in the Cartesian coordinate system XYZ.

Vector module and its orientation can be expressed as functions of its projections R_x, R_y, R_z as follows:

$$\begin{aligned} R &= \sqrt{R_x^2 + R_y^2 + R_z^2} \\ A_z &= \tan^{-1}\frac{R_x}{R_y} \\ El &= \tan^{-1}\frac{R_z}{\sqrt{R_x^2 + R_y^2}} \end{aligned} \tag{1}$$

It can be noted that measurement of referenced vectors can be used for the determination of satellite position or angular orientation. A minimum of three vectors is required to determine satellite position and two to determine its attitude. If more vectors are measured providing informational redundancy, then such statistical estimation methods as least square method (LSM) and Kalman filter (KF) can be applied. Satellite velocity and angular rate can be derived by the differentiation of its position and attitude applying a kind of filter recommended by the filtering and estimation theory [13–15]. It should also be noted that if vector orientation is measured for the position determination, then satellite attitude should be known and vice versa.

Especial autonomous satellite navigation system (sensor) is the inertial navigation system (INS/inertial measurement unit (IMU)). It can be used for the determination of satellite position, velocity, orientation and angular rate simultaneously. INS is based on measuring with linear accelerometers and angular rate sensors (“gyros”) the two vectors: satellite linear active acceleration $\overline{a}$ and angular rate $\overline{\omega}$. After integration, the system provides satellite position, velocity, attitude and angular rate. It is also assumed in INS theory that the vector of Earth gravity acceleration $\overline{g}$ is not measured by the system accelerometers, but it is computed from referenced mathematical Earth gravity field model. Essential INS disadvantage is that its errors grow with time. That is why, it has to be periodically corrected by such navigation aids as a pair of VMD used for the direct attitude determination. A detailed system description is out of this chapter’s scope and can be found in many publications [16–18]. Only the use of angular rate sensors ("gyros") for determination of satellite attitude is briefly considered below.

3.2.1.1 Determination of satellite position and velocity (GPS)

Today, satellite GPS can provide onboard accurate data about position, velocity and time [19] (**Figure 9**).

Accuracy: position, 15 m (2σ); velocity, 1.5 m/s (2σ); time, 1 μS.

Figure 9.
Satellite GPS SRG-10. Double redundant with a pair of zenith and nadir antennas.

GPS receiver is a radio range measuring device that measures distance from the desired satellite to navigation satellite constellation (NAVSTAR, USA; GLONASS, Russia; and GALILEO, Europe) and computes its position and velocity. GPS measures the distance R ($R = \sqrt{R_x^2 + R_y^2 + R_z^2}$) of the vector from the desired satellite to the navigation satellite, and this system is invariant of the system orientation (satellite attitude). The distance between the desired satellite and navigation satellite is measured by measuring the time delay Δt between the time t^s of the radio pulse transmitted by navigation satellite and the reception time $t^r$of its reception by GPS receiver installed on the desired satellite $\Delta t = t^r - t^s$. Measuring the distance allows to determine the desired satellite relative position (relatively to navigation satellite), and using known navigation satellite position that is continuously received by the receiver for every tracking satellite in the navigation message (NM) converts it in absolute position.

A minimum of three navigation satellites should be simultaneously traced by the receiver to determine position and velocity. Then satellite position is the cross-point of three spherical surfaces of the position equation $R^i = const, \quad i = 1, 2, 3$. If more tracked satellites are available, then redundant information can be used to calibrate the onboard clock (using the four satellites) and to use least square method or Kalman filter. Four nonlinear algebraic equations (pseudo-range measurements) are usually used to determine satellite position with GPS receiver:

$$R^i = \sqrt{(x - x^i)^2 + (y - y^i)^2 + (z - z^i)^2} + c\tau \qquad (2)$$
$$i = 1, 2, 3, 4$$

where R^i is the distance (pseudo-range) to the ith tracked navigation satellite, $x^i, y^i, z^i$are the navigation satellite Cartesian coordinates received in the NM, x, y, z are the Cartesian coordinates of the desired satellite, $c = 299\ 792\ 458\ km/s$is the speed velocity and τ is the GPS receiver clock bias. Satellite position can be found by solving it (1) numerically. It could also be linearized by using redundant measurements ($t > 4$) with LSM or KF. Satellite velocity can be determined by the differentiation of its position. Finally, GPS receiver can provide to AODC OBC current satellite position and velocity in the reference (e.g., Earth-centered inertial (ECI) [14] frame x, y, z, V_x, V_y, $V_z$and synchronized (by GPS) onboard time t^s).

3.2.1.2 Determination of satellite attitude and angular rate

3.2.1.2.1 TRIAD method (MAG, SS, HS)

The TRIAD method [10] is applied when two different vectors are measured. They usually can be any of the three pairs combined with the following three vectors: Earth magnetic induction vector $\overline{B}$ (measured with three-axis MAG), Sun vector $\overline{S}$ (measured with two-axis SS), and local vertical $\overline{r}$ (perpendicular to the local infrared radiation temperature surface, measured with the HS). At least two different not collinear vectors (their orientation) should be measured to determine satellite attitude that here is considered as satellite directional cosine matrix (DCM) and related three Euler angles of the certain order of rotations (e.g., 3-2-1) [9, 10] .

Let us assume that two different physical nature not collinear vectors $\overline{U} = \overline{S}$ and $\overline{V} = \overline{r}$ are measured $\overline{U}_m, \overline{V}_m$ by two vector measuring devises (SS and HS) installed on the satellite board and both these vectors are referenced in the reference frame as $\overline{U}_r, \overline{V}_r$. Let us choose $\overline{U}$ as the main vector and $\overline{V}$ as an auxiliary vector. Then an orthogonal coordinate system (frame) with basis unit vectors, $\overline{q}, \overline{r}$ and $\overline{s}$ can be defined as follows [10, 20]:

$$\vec{q} = \frac{\vec{U}}{|\vec{U}|}$$

$$\vec{r} = \frac{\vec{U} \times \vec{V}}{|\vec{U} \times \vec{V}|} \tag{3}$$

$$\vec{s} = \vec{q} \times \vec{r}$$

These unit vectors expressed at a given time by measured values in measured frame or body frame and reference values in a reference frame define two rotation matrixes, $\mathbf{C_m}$ and $\mathbf{C_r}$, as follows:

$$\begin{aligned} \mathbf{C_m} &= \left[\mathbf{q_m}; \mathbf{r_m}; \mathbf{s_m}\right] \\ \mathbf{C_r} &= \left[\mathbf{q_r};\ \ \mathbf{r_r};\ \ \mathbf{s_r}\right] \end{aligned} \tag{4}$$

where vectors $\overline{q}, \overline{r}, \overline{s}$ are written in the matrix form as matrix columns.

Rotation matrix C_{br} that defines attitude in the body frame with respect to reference frame is determined by the following formula:

$$\mathbf{C_{br}} = \mathbf{C_m} \cdot \mathbf{C_r} \tag{5}$$

Three Euler angles of rotation, roll (ϕ), pitch (θ), and yaw (ψ), can be expressed through the elements of the matrix C_{br}. Certain trigonometric formulas depend on the agreement about the order of the body rotations. For the order 3-2-1, the matrix C_{br} is as follows [9, 10]:

$$\mathbf{C_{br}} = \begin{bmatrix} c\theta c\psi & c\theta s\psi & -s\theta \\ s\phi s\theta c\psi - c\phi s\psi & s\phi s\theta s\psi + c\phi c\psi & s\phi c\theta \\ c\phi s\theta c\psi + s\phi s\psi & c\phi s\theta s\psi - s\phi c\psi & c\phi c\theta \end{bmatrix} \quad (6)$$

where c and s stand for cosine and sine angle. Then, formulas for Euler angles can be derived from (6) as:

$$\begin{aligned} \phi &= \tan^{-1}\frac{C_{23}}{C_{33}}, \\ \theta &= -\sin^{-1}C_{13}, \\ \psi &= \tan^{-1}\frac{C_{12}}{C_{11}} \end{aligned} \quad (7)$$

Vector measured sensors

If a pair from the three vectors (B, S, r-write as vectors) is measured, then following VMD in the pair can be used: SS (**Figure 10**), HS (**Figure 11**) and MAG (**Figure 12**).

Figure 10.
S-vector sensor Bradford fine sun sensor, accuracy, 0.2 deg. (2σ).

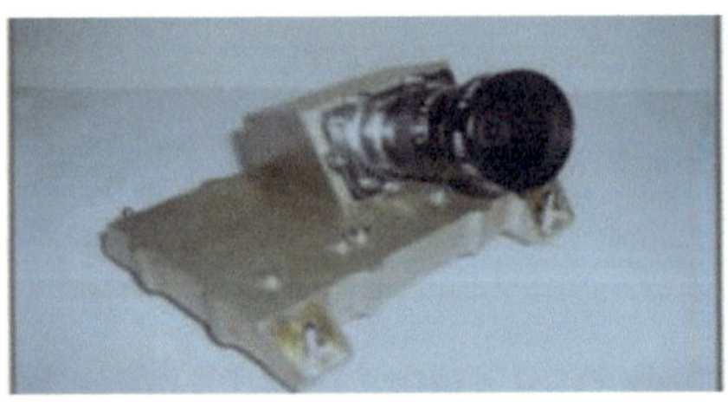

Figure 11.
r-vector sensor HS CMOS/SRAM-modular infrared horizon sensor, accuracy, 0.4 deg. (2σ).

Figure 12.
B-vector sensor MAG TFM100-S, accuracy, 10mG (2σ).

3.2.1.2.2 LSM method for star tracker (ST)

If more than two vectors are measured and available for attitude determination, then LSM-BATCH method [10] can be applied to use informational redundancy for increasing the stochastic estimation accuracy. This method basically can be applied for any set of VMD but is specifically convenient for the star tracker (ST), when some number (n) of navigation stars are in the device field of view (FOV) and are detected and tracked simultaneously, providing measured vectors $\overline{R}_m$ to these stars that are referenced in the device space catalog $\overline{R}_r$ (**Figures 10–13**).

Figure 13.
Star direction R-vector measured sensor (optic and computer units). Advanced stellar compass, accuracy, $2'' - 16''$ (2σ).

Let us consider the transformation of the referenced vector $\overline{R}_r$ in satellite body frame, where it is measured with the ST

$$\mathbf{R_m} = \mathbf{CR_r} \tag{8}$$

where $\mathbf{C}$ is the DCM of the rotation from the reference frame to satellite body frame and vectors $\overline{R}_r$ and $\overline{R}_m$ are written in the matrix form as matrix columns.

If the ST is in the tracking mode keeping in its FOV some n detected navigation stars, then it can be assumed that $\mathbf{C}$ is a small-angle matrix that is independent of the rotation order and can be expressed as follows:

$$\mathbf{C} \approx \begin{bmatrix} 1 & \alpha_z & -\alpha_y \\ -\alpha_z & 1 & \alpha_x \\ \alpha_y & -\alpha_x & 1 \end{bmatrix} \tag{9}$$

where $\alpha_x, \alpha_y, \alpha_z$ are small angles of satellite rotation about X, Y, Z axis, respectively.

Then subtracting from (8) $\mathbf{R_r}$, the following equation can be written:

$$\boldsymbol{\delta}\mathbf{R} = \boldsymbol{\delta}\mathbf{CR_r} \tag{10}$$

where $\boldsymbol{\delta}\mathbf{R} = \mathbf{R_m} - \mathbf{R_r}$, $\boldsymbol{\delta}\mathbf{C} \approx \begin{bmatrix} 0 & \alpha_z & -\alpha_y \\ -\alpha_z & 0 & \alpha_x \\ \alpha_y & -\alpha_x & 0 \end{bmatrix}$.

Transforming in (10) matrix product and taking into account random measurement errors, this equation can be represented in the following form:

$$\boldsymbol{\delta}\mathbf{R} = \mathbf{R_r}\boldsymbol{\delta}\mathbf{C} + \mathbf{V} \tag{11}$$

where $\boldsymbol{\delta}\mathbf{R} = \begin{bmatrix} \delta R_x \\ \delta R_y \\ \delta R_z \end{bmatrix}$, $\mathbf{R_r} = \begin{bmatrix} 0 & -R_{rz} & R_{ry} \\ R_{rz} & 0 & -R_{rx} \\ -R_{ry} & R_{rx} & 0 \end{bmatrix}$, $\boldsymbol{\delta}\mathbf{C} = \boldsymbol{\alpha} = \begin{bmatrix} \alpha_x \\ \alpha_y \\ \alpha_z \end{bmatrix}$, $\mathbf{V} = \begin{bmatrix} V_x \\ V_y \\ V_z \end{bmatrix}$,

V is random measurement error vector (considered as the white Gaussian noise, having covariance matrix $\mathbf{R} = r\mathbf{I}$).

Then this equation can be considered as a "standard" linear algebraic equation:

$$\begin{gathered}\mathbf{z_i} = \mathbf{h_i x} + \mathbf{V_i} \\ i = 1, 2, ..n - \text{number of measured vectors}\end{gathered} \tag{12}$$

where, $\mathbf{z_i} = \mathbf{\delta R_i}$, $\mathbf{h_i} = \mathbf{R_{ri}} = \begin{bmatrix} 0 & -R_{rz} & R_{ry} \\ R_{rz} & 0 & -R_{rx} \\ -R_{ry} & R_{rx} & 0 \end{bmatrix}_i$, $\mathbf{V}_i = \begin{bmatrix} V_x \\ V_y \\ V_z \end{bmatrix}_i$ $\mathbf{x} = \mathbf{\delta C} = \boldsymbol{\alpha} = \begin{bmatrix} \alpha_x \\ \alpha_y \\ \alpha_z \end{bmatrix}$.

If $n = 1$, only a single vector is measured, then $\det\mathbf{h_i} = -R_{rx}R_{ry}R_{rz} + R_{rx}R_{ry}R_{rz} \equiv 0$, and thus, determining all three angles of satellite attitude is impossible. If it takes informational redundancy, then optimal estimate (providing minimum of standard deviation of satellite attitude errors) can be found with the following LSM formula [13]:

$$\hat{\boldsymbol{\alpha}} = \mathbf{Kz} \tag{13}$$

where $K = \left(H^T R^{-1} H\right)^{-1} H^T R^{-1} H, H = [h_1\, h_2\, h_3..h_n]^T$.

3.2.1.2.3 Determination of the angular rate

Direct measurement

Satellite angular rate $\overline{\omega}$ (vector absolute angular velocity) can be measured directly by three-axis rate sensor (RS) that could have mechanical, optical, or microelectromechanical systems (MEMS) design [21, 22] (**Figure 14**). Traditionally, independent of the design type, these RS are usually named "gyros," paying respect to their historical appearance for aerospace vehicle control purposes as a mechanical gyroscope (**Figure 15**).

Measured angular velocity vector $\overline{\omega}$ can be used to determine satellite attitude by the integration of matrix kinematic Poisson's Equation [9, 10]:

$$\dot{\mathbf{C}} = \breve{\boldsymbol{\omega}}\mathbf{C}, \mathbf{C}(\mathbf{0}) = \mathbf{C_0} \tag{14}$$

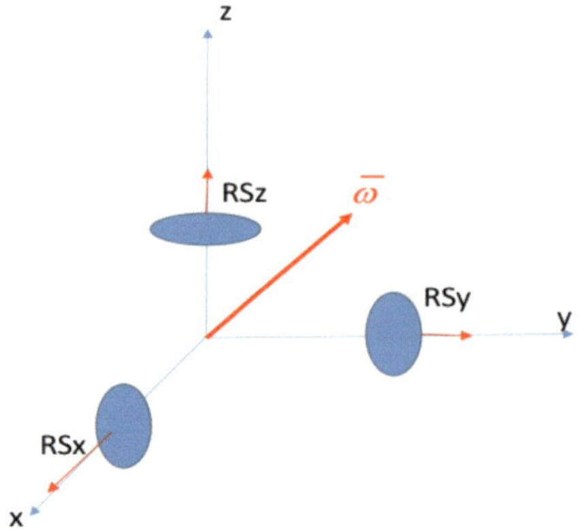

Figure 14.
Measurement of satellite angular velocity $\overline{\omega}$ with three rate sensors RS_x, RS_y, RS_z.

Figure 15.
ϖ-vector sensor RS, BEI QRS-11 single-axis body rate sensor, accuracy, $7\text{deg}/h = 0.0019\text{deg}/s\ (2\sigma)$.

where $C = C_{ib}$ $\mathbf{C} = \mathbf{C_{ib}}$ is the DCM between the inertial frame ECI and satellite body frame and $\breve{\omega} = \begin{bmatrix} 0 & \omega_z & -\omega_y \\ -\omega_z & 0 & \omega_x \\ \omega_y & -\omega_x & 0 \end{bmatrix}$ is skew symmetric matrix measured in the satellite frame components of vector of satellite absolute angular velocity. After determination of the DCM, satellite attitude in three Euler angles can be derived with Eq. (7) above. Unfortunately, gyro drift causes unlimited growing up errors in integrated attitude that require periodic corrections from two VMD, measuring the attitude directly (**Figures 10–12**).

Body rate estimator

Often, specifically for attitude stabilization (keeping or aka pointing) mode, satellite angular rate is estimated by using the so-called body rate estimator and is not measured directly by the RS. Indeed, using for attitude keeping mode small angles and linear approximation, we can simplify satellite attitude dynamics model [9] to three single-axis state equations and present it with the stochastic influences as follows:

$$\begin{cases} \dot{\boldsymbol{\omega}} = \mathbf{w} \\ \dot{\boldsymbol{\alpha}} = \boldsymbol{\omega} \\ \mathbf{z} = \boldsymbol{\alpha} + \mathbf{v} \end{cases} \tag{15}$$

where $\boldsymbol{\omega}$ is the angular velocity, $\boldsymbol{\alpha}$ is the satellite deviation angle from the desired direction and $\mathbf{z}$ is the satellite deviation angle measurement with random Gaussian white noise error $\mathbf{v}$. Realistically, it is a wide spectrum correlated process that components have spectral density $r_i = 2\sigma_{v_i}^2 T_{v_i}, i = x, y, z$ (σ_{v_i} is standard deviation of the random error v_i, T_{v_i} is $v_i(t)$ correlation time), $w_i$is exciting angular acceleration noise with spectral density $q_i = 2\sigma_{wi}^2 T_{wi}$ (σ_{w_i} is standard deviation of the random angular acceleration $w_i = \frac{M_i}{J_i}$, M_i is exciting external random torque, J_i is satellite moment of inertia, T_{w_i} is $w_i(t)$ correlation time).

The linear KF can be applied to synthesize the estimator for the optimal estimation of the vector angle $\boldsymbol{\alpha}$ and the vector of angular velocity $\boldsymbol{\omega}$, using noisy measurements $\mathbf{z}$ [9]:

$$\begin{cases} \dot{\hat{\boldsymbol{\omega}}} = \mathbf{k_{12}}(\mathbf{z} - \hat{\boldsymbol{\alpha}}) \\ \dot{\hat{\boldsymbol{\alpha}}} = \hat{\boldsymbol{\omega}} + \mathbf{k_{22}}(\mathbf{z} - \hat{\boldsymbol{\alpha}}) \end{cases} \tag{16}$$

where $\hat{\boldsymbol{\alpha}}$ and $\hat{\boldsymbol{\omega}}$ are the optimal estimates of the angle $\boldsymbol{\alpha}$ and the angular velocity $\boldsymbol{\omega}$ correspondingly.

Matrix KF (16) is separated in three independent scalar channels for X, Y, Z axis. Its weight coefficients $\mathbf{k}_{12}$ and $\mathbf{k}_{22}$ can be determined by solving KF Riccati Equation [13–15] for each of these three separate channels independently.

It can be shown that in the considering case, the steady-state ($t \rightarrow \infty$) KF coefficients are determined by the following formulas (identically for X, Y, Z axes, i = 1,2,3):

$$\begin{cases} k_{12_i} = \sqrt{\xi_i} \\ k_{22_i} = \sqrt{2\xi_i\sqrt{\xi_i}} \end{cases} \tag{17}$$

where $\xi_i = \frac{q_i}{r_i}$ is the ratio of spectral densities of satellite disturbing torque noise to measured attitude error noise (assuming that both are white Gaussian noises). This parameter can be considered as the *filterability index*. Eq. (16) can be represented in the transfer function (Laplace operator) form as a second-order differential equation unit:

$$\begin{cases} \hat{\omega}_i = \dfrac{s}{T_i^2 s^2 + 2d_i T_i s + 1} z_i \\ \\ \hat{\alpha} = \dfrac{2dT_i s + 1}{T_i^2 s^2 + 2d_i T_i s + 1} z_i \end{cases} \tag{18}$$

where s is the Laplace operator, T_i is the time constant, and d_i is the specific damping coefficient that is determined by the following formulas:

$$\begin{cases} T_i = \dfrac{1}{\sqrt{k_{12_i}}} = \dfrac{1}{\sqrt[4]{\xi_i}} \\ \\ d_i = \dfrac{k_{22_i}}{2\sqrt{k_{12i}}} = \dfrac{\sqrt{2}}{2} = 0.707 \end{cases} \tag{19}$$

or in other words, the time constant is in inverse proportionality to the filterability index (in ¼ degree) and the specific damping coefficient is conventional for such a second-order unit 0.707 for each of the three channels.

3.2.1.2.4 Multisensory sensor unit (MSU)

As it can be seen from the consideration above, the use of directly measuring devices (e.g., ST and RS) for attitude and body rate determination has a disadvantage. The random noises are at the devices output, and they have to be filtered in the closed control loop of satellite attitude control that puts some constraints to choose the control law coefficients. However, using indirect body rate measurement, the state estimator (filter) unavoidably introduces additional phase delay in the control loop because of the consecutive inclusion of this filter in the control loop. To use the RS (gyro) and the integrator for body rate and attitude determination autonomously for a long time is not possible because of the accumulated attitude errors caused by the integration of the gyro drift. The following scheme (that is common in Aviation) can be considered as free from the disadvantages above. Let us assume that satellite attitude is determined in two ways: continuous integration of RS angular velocity (IMU) and using VMD, for example, ST. Then this ST is used to correct the attitude derived by the integration of RS output. The idea of MSU is shown in **Figure 16**.

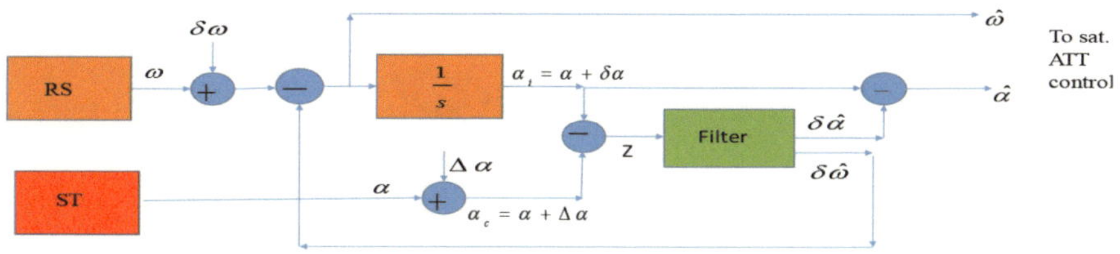

Figure 16.
Integration of multisensory sensor unit (MSU) single-axis channel.

In integrated IMU attitude (IMU = RS + integrator) as in **Figure 14** above (three identical channels), α_i (inertial angle) has been growing with time deterministic error $\delta\alpha$ due to the integration of RS bias $\delta\omega$ and ST has random noisy error $\Delta\alpha\Delta\alpha$. The difference of these signals is equal to the difference of system errors $z = \delta\alpha - \Delta\alpha$. This difference is used for estimation of errors of IMU with a kind of filter, and after the compensation at the system output, the estimates ($\hat{\alpha}$ and $\hat{\omega}$) can be used for satellite control. As it can be seen, in this scheme the filter is not connected in the control loop, and consequently, it does not introduce additional phase delay; however, the scheme still performs its job to filter the noise and estimate RS bias. This scheme can be very effective in practice. It can be presented similar to Eqs. (15) and (16) as follows:

$$\text{IMU model equations}: \begin{cases} \delta\dot{\omega} = w_1 \\ \delta\dot{\alpha} = \delta\omega \\ z = \delta\alpha + \Delta\alpha \end{cases} \tag{20}$$

where $\delta\omega$ is the RS error of measuring the angular velocity; $\delta\alpha$ is attitude error after the integration of the RS signal; $z$is IMU attitude error measurement with random Gaussian white noise error v $\Delta\alpha = v$, having spectral density $r_1 = 2\sigma^2_{\Delta\alpha}T_{\Delta\alpha}$ ($\sigma_{\Delta\alpha}$ is the standard deviation of the random error $\Delta\alpha$, $T_{\Delta\alpha}$ is its correlation time); and w_1 is the exciting noise of RS random drift with spectral density $q_1 = 2\sigma^2_{w1}T_{w1}$ (σ_{w1} is the standard deviation of the random drift, T_w is the $w(t)$ correlation time).

$$\text{KF equations}: \begin{cases} \delta\dot{\hat{\omega}} = k_{12}(z - \delta\hat{\alpha}) \\ \delta\dot{\hat{\alpha}} = \delta\hat{\omega} + k_{22}(z - \delta\hat{\alpha}) \end{cases} \tag{21}$$

where KF coefficients k_{12} and k_{22} are determined by (17), substituting there q_1 and r_1 instead of q and r.

3.2.2 Actuators

3.2.2.1 Propulsion system (PS)

Satellite propulsion system [9, 10] is usually designed for satellite orbital and/or angular control. In the first case, PS is commanded from the ground OC by TLM commands in some cases when satellite orbit has to be changed (orbit correction, deorbiting, collision avoidance), in the second controlled automatically from onboard AODCS. It consists of such typical elements as orbital and attitude thrusters (number and installation scheme depending on certain application), propulsion tank with associated pipes, valves, regulators, and electronics. General principles of PS act independently of the type (ion thrusters (0.01–0.1 N), liquid propellant and solid motor (100–10,000 N), cold gas (1-3 N)).

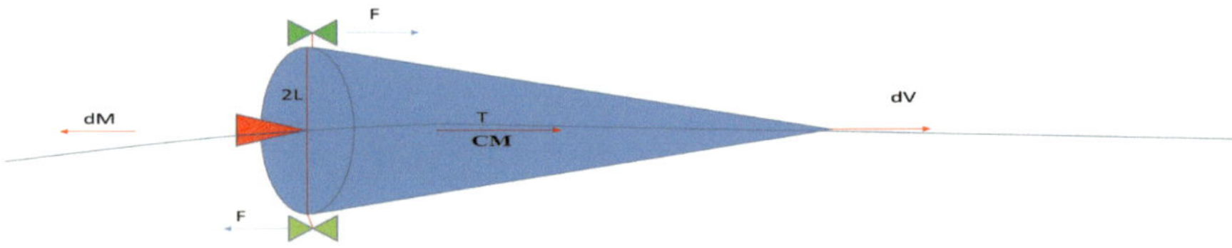

Figure 17.
Satellite control with PS thruster principles.

Figure 15 illustrates the satellite control with PS thruster principles. The principle of the formation of the propulsion jet force can be presented by the following equation of variable mass body dynamics that from Russian sources, for example, [23], is known as Prof. I. Meshchersky's equation:

$$M\frac{dV}{dt} = F - V_p\frac{dm}{dt} \tag{22}$$

where M is the mass body, F is the external force, V_P is the propellant exhaust velocity and m is the propellant mass. The term $-V_p\frac{dm}{dt}$ is the propulsion force (propulsion thrust) (**Figure 17**).

$$T = -V_p\frac{dm}{dt} \tag{23}$$

In Section 3.2.2.1.2, it is always $\frac{dm}{dt} < 0$, then $T > 0$. Usually for any propulsion system, let us introduce parameter *specific pulse* I_{sp}, where $I_{sp} = \frac{T}{g|\dot{m}|} = \frac{V_p}{g}$ then $V_p = I_{sp}g\dot{m}$ and $T = -I_{sp}g\dot{m}$. If the satellite thruster is installed such as satellite point of the center of mass (CM) is located on the line of the action of the force T (red thruster in **Figure 17**), then the thruster can serve for satellite orbit correction, and the pulse of the control thrust causes increment of satellite velocity $dV = \frac{1}{M}Tdt$. If satellite thrusters are installed in such a way that having an arm L from the CM, then they create rotating torque (green reversible thruster pair in **Figure 17**) and can be used for attitude control (control torque is $T_r = LT$). And the increment in attitude angle for the time dt of the pair of thruster activation will be $d\alpha = \frac{T_r}{J}dt$.

The expelled propulsion mass Δm can be calculated with the K. Tsiolkovsky formula [9, 24] that follows from Eq. (22)

$$\Delta m = m_0\left[1 - e^{-\frac{dV}{gI_{sp}}}\right] \tag{24}$$

where $m_0$is the propellant initial mass. Photos of I. Meshchersky and K. Tsiolkovsky are presented (**Figures 18** and **19**).

Discrete pulse modulation control is usually used to minimize the consumption of the propellant for attitude control [9]. Examples of the gas thruster and the tank are presented in **Figures 20** and **21**.

3.2.2.2 Magnetic torque rods (MTR)

Magnetorquers are essentially sets of electromagnets. A conductive wire is wrapped around a ferromagnetic core which is magnetized when excited by the electric current caused by the control voltage applied to the coil. The disadvantage of this design is the presence of a residual magnetic dipole that remains even when

Figure 18.
Prof. I. Meshchersky (1859–1935).

Figure 19.
K. Tsiolkovsky (1857–1935).

Figure 20.
Cold gas GN-2 thruster, nominal thrust 3.6 N (230 psi), specific impulse 57 s.

the coil is turned off because of the hysteresis in the magnetization curve of the core. It is therefore necessary to demagnetize the core with a proper demagnetizing procedure. Normally, the presence of the core (generally consisting of ferromagnetic) increases the mass of the system. The control voltage is controlled by AODCS control output (**Figures 18–20**). The magnetic dipole generated by the magnetorquer is expressed by the formula:

$$\overline{M} = ni\overline{S} \tag{25}$$

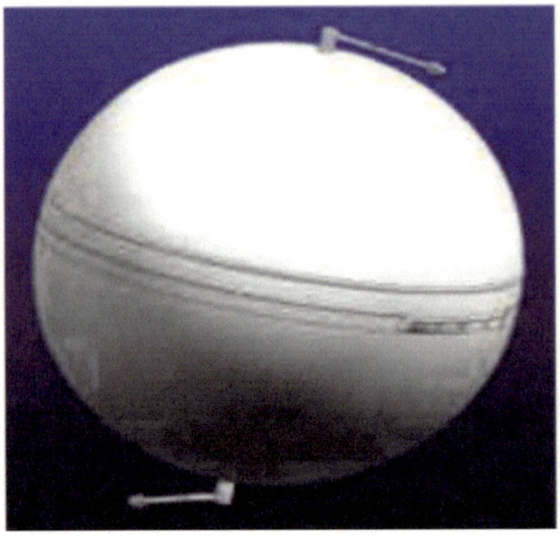

Figure 21.
60 liter propulsion gas tank.

where n is the number of turns of the wire, i is the current provided, and $\overline{S}$ is the vector aria of the coil. The dipole interacts with the Earth magnetic field, generating a torque whose expression is:

$$\overline{T}_m = \overline{M} \times \overline{B} \tag{26}$$

where $\overline{M}$ is the magnetic dipole vector moment, $\overline{B}$ is the Earth magnetic field induction vector and $\overline{T}_m$ is the generated magnetic torque vector. This equation in the scalar form is as follows:

$$\begin{bmatrix} T_{mx} \\ T_{my} \\ T_{mz} \end{bmatrix} = \begin{bmatrix} 0 & B_z & -B_y \\ -B_z & 0 & B_x \\ B_y & -B_x & 0 \end{bmatrix} \begin{bmatrix} M_x \\ M_y \\ M_z \end{bmatrix} \tag{27}$$

Typically, three coils are used; the three-coil assembly usually takes the form of three perpendicular coils, because this setup equalizes the rotational symmetry of the fields which can be generated; no matter how the external field and the craft are placed with respect to each other, approximately the same torque can always be generated simply by using different amounts of current on the three different coils (**Figure 22**).

It can be seen from Eq. (26) that MTR cannot generate the magnetic torque in the direction that is parallel to Earth magnetic field $\overline{B}$ ($\overline{M} \| \overline{B}$) and it always is perpendicular to the Earth magnetic field vector $\overline{T}_m \perp \overline{B}$. Unfortunately, from Eq. (27), the required magnetic moments cannot be found, because it has zero determinant, and we cannot invert it.

$$\Delta = \begin{vmatrix} 0 & B_z & -B_y \\ -B_z & 0 & B_x \\ B_y & -B_x & 0 \end{vmatrix} = -B_x B_y B_z + B_x B_y B_z = 0 \tag{28}$$

However, the following approach can be used to find required vector $\overline{M}$ [9]. When we take cross-product of $\overline{B}$ with both sides of Eq. (26) and take into account that it is useless to apply $\overline{M}$ in parallel direction to $\overline{B}$, and hence we can require that $\overline{M} \perp \overline{B}$ and $\overline{M} \cdot \overline{B} = 0$, then the following formula can be derived:

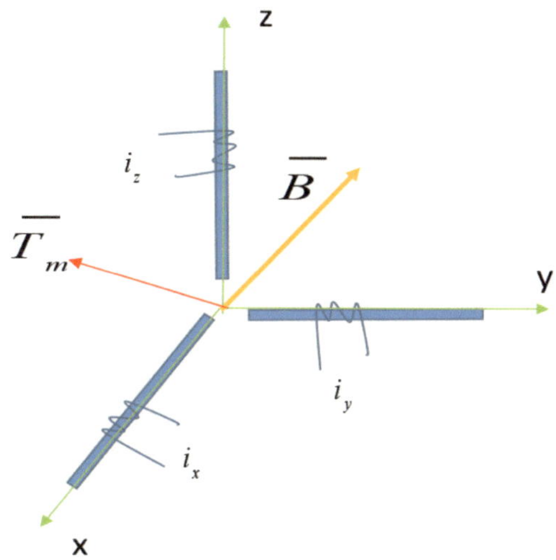

Figure 22.
3D orthogonal magnetic torque rods.

$$\overline{M} = \frac{\overline{B} \times \overline{T}_m}{B^2} \tag{29}$$

Another MTR control method is the so-called B-dot control [25].

$$\overline{M} = -k\dot{\overline{B}} \tag{30}$$

where k is control gate coefficient.

As the result of (29) control, the satellite will reduce its body rate and is finally slow rotated along Earth's geomagnetic field line (vector $\overline{B}$). Eventually it achieves its threshold of capture by the gravity gradient effect.

If the redundancy is required, it is provided by additional (redundant) coil with the same core (**Figure 23**).

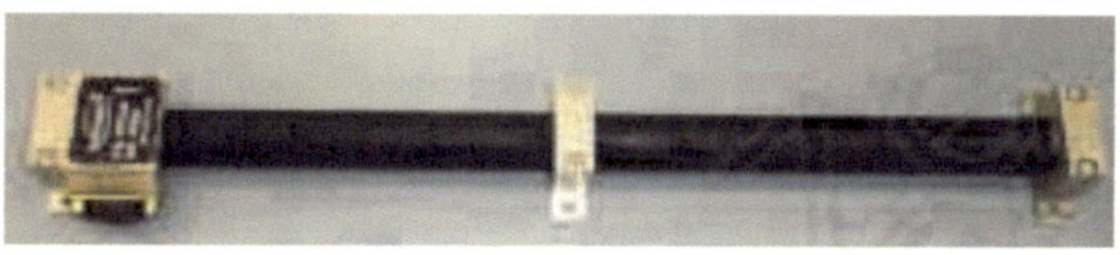

Figure 23.
Magnetic torque rod SSTL MTR-30, magnetic moment, M = 30 Am².

3.2.2.3 Reaction/reaction-momentum wheels (RW/RMW)

Reaction wheels (RW), aka momentum exchange devices [9] or reaction-momentum wheels (RMW), have massive rotated rotor with big axial moment of inertia with respect to the axis of rotation. They are electrically controlled by the electric motors and the rotor is installed on the rotating motor shaft. The controlled voltage, applied to the control winding of the motor, controls its rotor angular speed. The product of the rotor angular acceleration $\dot{\mathbf{\Omega}}$ multiplied by its axial inertia $\mathbf{I}$ is the RW generated inertia control torque $\mathbf{T_i} = \mathbf{I}\dot{\mathbf{\Omega}}$ that is applied to the satellite body in opposite to the acceleration direction. Special embedded angular speed

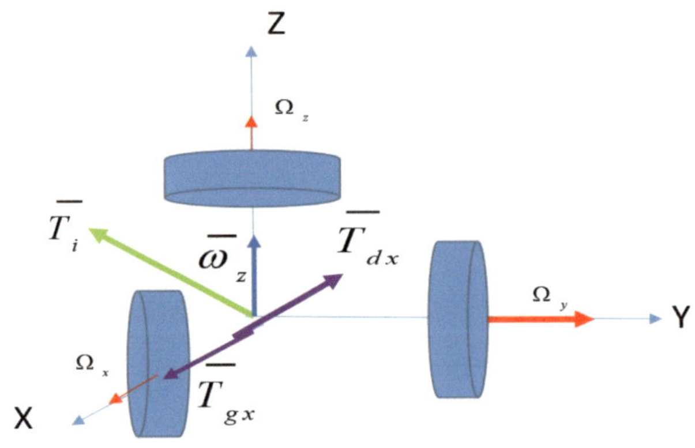

Figure 24.
Three orthogonal reaction wheels (RW).

sensor (tachometer) measures the motor rotor angular speed and allows organizing RW inertia torque control proportional to the applied control voltage. At least three RW, as in **Figure 24**, are required to produce control torque vector in three-dimensional spatial, having desired value and pointed in the desired direction. Sometimes redundant unit of three redundant wheels or one with 4th redundant skewed wheel is applied to meet reliability requirements.

RW can generate control inertia torques $\mathbf{T_i}$ only when they are accelerated or decelerated. With this torque, they cannot compensate a permanent disturbance torque $\mathbf{T_d} = \mathbf{const}$ applied to satellite for a long enough time and come eventually to some maximum/minimum available speed aka *the saturation speed*. At that state $\overline{T}_i$ becomes zero. That is why RW is usually applied with other types of actuators such as MTRs or gas thrusters to de-saturate the RW to use them as the source of $\mathbf{T_i}$ again.

In general case, RW can be run around in some nominal angular speed $\mathbf{\Omega_0}$. In this case they can be named as reaction-momentum wheel (RMW) and then just RW. Mathematically RW/RMW dynamics can be presented as follows. Let us consider satellite with RW unit angular momentum:

$$\mathbf{H_s} = \mathbf{H} + \mathbf{h} \tag{31}$$

where $\mathbf{H} = \mathbf{J}\boldsymbol{\omega}$ is the satellite *absolute* angular momentum vector (column matrix), $\boldsymbol{\omega}$ is the vector of satellite absolute angular velocity, $\mathbf{J} = \begin{bmatrix} J_{xx} & -J_{xy} & -J_{xz} \\ -J_{yx} & J_{yy} & -J_{yz} \\ -J_{zx} & -J_{zy} & J_{zz} \end{bmatrix}$ is the satellite inertia matrix, $\mathbf{h} = \mathbf{I}\boldsymbol{\Omega}$ is the RMW *relative* angular momentum vector, $\boldsymbol{\Omega}$ is the RW relative rotation speed vector, and $\mathbf{I} = \begin{bmatrix} I_{xx} & 0 & 0 \\ 0 & I_{yy} & 0 \\ 0 & 0 & I_{zz} \end{bmatrix}$ is the RW inertia matrix.

Then differentiating (31) in rotating with angular velocity satellite axis $\boldsymbol{\omega}$ and using Euler's rigid body dynamics formula [9], we can get the following equation:

$$\dot{\mathbf{H}} + \breve{\boldsymbol{\omega}}\mathbf{H} = -\dot{\mathbf{h}} - \breve{\boldsymbol{\omega}}\mathbf{h} + \mathbf{T} \tag{32}$$

where $\breve{\boldsymbol{\omega}} = \begin{bmatrix} 0 & -\omega_z & \omega_y \\ \omega_z & 0 & -\omega_x \\ -\omega_y & \omega_x & 0 \end{bmatrix}$ is the satellite angular velocity matrix and $\mathbf{T}$ is the vector of the external torque applied to the satellite. In the right side of Eq. (32),

we can see two terms that have meaning of torques applied from the RMW unit to the satellite body: $-\dot{\mathbf{h}} = \mathbf{T_i}$ the inertial torque and $-\breve{\boldsymbol{\omega}}\mathbf{h} = \mathbf{T_g}$ the gyro torque. They are RMW generated torques that can be used for the satellite attitude control.
Eq. (32) presents satellite attitude dynamics under the action of RMW torques. The RW dynamics can be presented similar to Eq. (32):

$$\dot{\mathbf{h}}_{\mathbf{w}} + \breve{\boldsymbol{\omega}}\mathbf{h}_{\mathbf{w}} = \mathbf{M}_{\mathbf{w}} + \mathbf{M}_{\mathbf{f}} \tag{33}$$

where $\mathbf{h_w} = \mathbf{h} + \mathbf{h_c}$ is the absolute RW momentum, $\mathbf{h_c} = \mathbf{I}\boldsymbol{\omega}$ is the carrier component of RW absolute momentum, **QUOTE** $\mathbf{M_w}$is the RW motor control torque, and $\mathbf{M_f}$ is the RW friction torque.

Eq. (33) can be represented in the following form:

$$\dot{\mathbf{h}} = \mathbf{I}\dot{\boldsymbol{\Omega}} = -\mathbf{I}\dot{\boldsymbol{\omega}} - \breve{\boldsymbol{\omega}}(\mathbf{h} + \mathbf{h_c}) + \mathbf{M_w} + \mathbf{M_f} \tag{34}$$

where $\mathbf{M_f} = -\mathbf{k_e}\boldsymbol{\Omega} - \mathbf{M_{df}}\,\mathrm{sgn}\,\boldsymbol{\Omega}$ is the friction torque, $\mathbf{k_e}$ is the motor natural damping coefficient, $\mathbf{M_{df}}$ is the dry friction torque, and $\mathbf{M_w}$ is the control motor torque.

The torque $\mathbf{M_f}$ usually consists of two components: the viscous torque (counter electromotive voltage in the control coil) $-\mathbf{k_e}\boldsymbol{\Omega}$ and dry or Coulomb friction torque in the RMW bearings—$\mathbf{M_{df}}\,\mathrm{sgn}\,\boldsymbol{\Omega}$. The control torque $\mathbf{M_w}$ can be set as follows:

$$\mathbf{M_w} = -(\mathbf{k_\Omega} + \mathbf{k_e})(\boldsymbol{\Omega} - \boldsymbol{\Omega_0}) - \mathbf{k_T}\mathbf{T_c} \tag{35}$$

where $\mathbf{k_\Omega}$ is the motor control damping coefficient, $\boldsymbol{\Omega_0}$ is the desired angular RMW speed, $\mathbf{k_T}$ is the control torque coefficient, and $\mathbf{T_c}$ is requested from AODCS OBC control torque. Taking into account Eq. (35), Eq. (34) can be rewritten in the following form:

$$\mathbf{I}\dot{\boldsymbol{\Omega}} + (\mathbf{k_\Omega} + \mathbf{k_e})\boldsymbol{\Omega} = (\mathbf{k_\Omega} + \mathbf{k_e})\boldsymbol{\Omega_0} + \mathbf{T_c} - \mathbf{I}\dot{\boldsymbol{\omega}} - \breve{\boldsymbol{\omega}}(\mathbf{h} + \mathbf{h_c}) + \mathbf{M_f} \tag{36}$$

In the operator Laplace s-form (transfer function), Eq. (36) can be rewritten as follows:

$$\boldsymbol{\Omega}(s) = \mathbf{W}(s)\left\{\boldsymbol{\Omega_0} + \mathbf{k}_w^{-1}\left[\mathbf{T_c} - \mathbf{I}s\boldsymbol{\omega} - \breve{\boldsymbol{\omega}}(\mathbf{h} + \mathbf{h_c}) + \mathbf{M_f}\right]\right\} \tag{37}$$

where $\mathbf{W}(s) = \begin{bmatrix} (T_{wx}+1)^{-1} & 0 & 0 \\ 0 & (T_{wy}+1)^{-1} & 0 \\ 0 & 0 & (T_{wz}+1)^{-1} \end{bmatrix}$ is the RW matrix transfer function, $T_{w_x} = \frac{I_x}{k_{\Omega x}+k_{ex}}, T_{w_y} = \frac{I_x}{k_{\Omega y}+k_{ey}}, T_{w_x} = \frac{I_x}{k_{\Omega z}+k_{ez}}$ is the RW time constants, and.

$\mathbf{k}_w = \begin{bmatrix} (k_{\Omega x}+k_{ex})^{-1} & 0 & 0 \\ 0 & (k_{\Omega y}+k_{ey})^{-1} & 0 \\ 0 & 0 & (k_{\Omega z}+k_{ez})^{-1} \end{bmatrix}$ is the RW control speed coefficient.

It should be noted that sometimes the RW control loop is more sophisticated. Special integrators could be connected into the loop to memorize and compensate the dry friction torques acting in bearings. Some small nominal rotating speed can be set for all three RMW to eliminate the dry friction torque having a pike when the wheel speed is zero $\boldsymbol{\Omega} = \mathbf{0}$.

However, more representative case is when $\mathbf{\Omega_0}$ is set (usually in one direction as in **Figure 23**—axis Y) to provide to the satellite a gyro-stiffness (gyro-stabilization capability) with respect to another two axes X and Z.

Indeed, if we put in (32) that $\dot{\omega}_x = \dot{\omega}_y = \dot{\omega}_z = \dot{\Omega}_x = \dot{\Omega}_y = \dot{\Omega}_z = 0$, and $\omega_x = \omega_z = \Omega_x = \dot{\Omega}_y = \dot{\Omega}_z = 0, \quad \omega_y = \omega_0 = const, \Omega_y = \Omega_0 = const$, then it can be written as:

$$\begin{cases} -\omega_z (H_y + h_y) = T_x \\ \omega_x (H_y + h_y) = T_z \end{cases} \tag{38}$$

where $H_y = J_y \omega_0, h_y = I_y \Omega_0$. In this case carrier orbital angular velocity ω_0 helps to increase satellite momentum $H_{sy} = H_y + h_y$. Eq. (38) is known as three degrees of freedom of a free gyroscope precession [21] and represents the gyro-stabilization effect: that gyroscope vector H_{sy} being free of disturbing torques can keep its direction and value in inertial spatial (**Figure 25**).

Figure 25.
Reaction/momentum wheel HR-0610, torque, $75 \cdot 10^{-3}$Nm; *momentum,* $(4 - 12)$ Nms.

3.2.3 AODCS OBCS

Independently of sytem arhitecture; it is separate dedicated to AODCS computer, or a special AODCS card within central satellite C&DH computer, it is the integration element of AODCS [1, 11]. AODCS system may consist of the computer (computer card) itself (OBC) and auxiliary intercommunication electronic units (electronic cards) AEU carrying DC/DC electric power conversion and I/O (analog and digital) interface and commutation functions.

OBC can be divided into two parts: the hardware (HW, power convertor, processor, input/output [I/O] convertors, non-volatile and volatile memory) and the software (SW, operation system [OS] and vital or functional software [VS/FS]) (**Figure 26**).

What makes the satellite OBC essentially different for the airplane OBC is that its SW can be uploaded and updated from the ground and during operation and scheduled maintenance. OS OBC includes generic computer programs: program of I/O interface, time schedule (dispatcher), embedded test, timer and standard mathematic functions. Satellite SW often is considered as satellite *SW subsystem* that is verified during development (with mathematical high-fidelity Matlab/Simulink simulators and semi-natural processor-in-the-loop (PIL) simulators). SW subsystem should be tested to meet SW requirements [26, 27]. The flight version of the SW subsystem is supported with operation real-time satellite simulators (RSS) [1, 11] located in operation center. It should be mentioned that only final AODCS

Figure 26.
Satellite OBCS, MAC-200 (C&DH unit with AODCS card) comprises of two OBC: Prime and redundant (cold reserve).

(OBC (HW + SW), sensors, and actuators) *functional test* [2] that should be performed in the Space Qualification Laboratory [7] during satellite Space Qualification and Acceptance campaign can really minimize the risk of launching a not ready satellite and prevent against AODCS refinishing in orbit during commissioning and operation.

VS can be separated in two parts, ODCS SW and ADCS SW. For both parts, I/O interface with sensors and actuators is determined in special interface control document(s) (ICD), describing type, certain connectors, and electrical parameters of the exchanging data. These data before using them for functional tasks are pre-processed in OBC with special algorithms.

3.2.3.1 Satellite sensors/actuators data preprocessing

This group of algorithms performs the following common tasks:

- Convert data into required physical parameters and units, taking into account certain sensor input–output scale function.
- Transform data in certain device frame and compensate device misalignment, bias and scale function errors if it is possible, monitor device state, establishing "on/off," "work/control," "data bad/good" flags.
- Transfer to C&DH TLM data about sensor/actuator state and their data.
- Perform some other auxiliary functions if they are required.
- Main functional tasks ODCS SW and ADCS SW can be listed as below.

3.2.3.2 ODCS SW

3.2.3.2.1 Satellite orbit propagation (OP)

To understand the idea of propagation of satellite orbit in Earth gravity field to the simplest, Keplerian motion propagator based on spherical Earth gravity field model might be used [9]; however more realistic results can be obtained with more accurate propagator, taking into account the second zonal harmonic J_2 in the function of approximation of Earth gravitational potential. The following equations of motion of satellite center of mass in Earth gravitational field can be considered [9]:

$$\ddot{x} = -\mu\frac{x}{r^3} + A_{J2}\left(15\frac{xz^2}{r^7} - 3\frac{x}{r^5}\right),$$
$$\ddot{y} = -\mu\frac{y}{r^3} + A_{J2}\left(15\frac{yz^2}{r^7} - 3\frac{y}{r^5}\right), \tag{39}$$
$$\ddot{z} = -\mu\frac{z}{r^3} + A_{J2}\left(15\frac{z^3}{r^7} - 9\frac{z}{r^5}\right),$$
$$r = \sqrt{x^2 + y^2 + z^2},$$

where x, y, z are the Cartesian coordinates of satellite center of mass in inertial frame ECI, $r = \sqrt{x^2 + y^2 + z^2}$ is the module of the radius vector from the center of Earth to satellite center of mass, $\mu = 3.986004418 \cdot 10^{14}$ $[m^3/s^2]$ is the Earth gravitational constant, $A_{J2} = \frac{1}{2}J_2 \cdot R_e^2$ is a constant, J_2 =0.00108263 is the second zonal harmonic coefficient in the raw of Earth potential function, and $R_e = 6378137.00\ m$ is the mean radius of the Earth at the equator.

These equations can propagate satellite position and velocity (x, y, z and $\dot{x}, \dot{y}, \dot{z}$) in the inertial Cartesian ECI coordinate system if the initial parameters are initially set $x_0 = x(0), y_0 = y(0), z_0 = z(0)$ and $\dot{x}_0 = \dot{x}(0), \dot{y}_0 = \dot{y}(0), \dot{z}_0 = \dot{z}(0)$. They can be periodically determined from GPS or MCC TLM information. The propagation credibility time depends on orbit perturbations [9, 10, 27] and required accuracy. The most accurate and common ground propagators are NORAD Simplified Perturbation Model (SGP) propagators. NORAD SGP is used for proving to users two-line element (TLE) satellite orbital data. For the low Earth orbit (LEO), having altitude below 6000 km (period about 225 min), they provide position accuracy about 1 km within a few days that for many users is accurate enough and needs to be updated once or twice per week. Currently almost every satellite is equipped with GPS and its onboard propagators are practically continuously corrected with GPS (and sometimes MCC TLM) data that provide position within 10–100 m and velocity within 0.01–0.1 m/s accuracy range. Only some short periods of GPS data outage require orbit propagation. In addition to satellite position and velocity, OP calculates conventional orbital parameters (**Figure 14**) that can be computed with the following formulas [9]:

$$a = \frac{\mu}{2\left[\frac{\mu}{r} - \frac{V^2}{2}\right]}$$
$$i = \cos^{-1}\frac{h_{zi}}{h}$$
$$\Omega = \tan^{-1}\frac{h_{xi}}{-h_{yi}} \tag{40}$$
$$\nu = \tan^{-1}\frac{\overline{r} \cdot \overline{V}}{p - r}$$
$$u = \sin^{-1}\frac{z_i}{r \sin i}$$
$$\omega = u\text{-}\theta$$

where $\overline{h} = \overline{r} \times \overline{V}$ is the satellite orbital linear momentum vector, a is the satellite orbit semi-major axis, $h = \sqrt{h_{xi}^2 + h_{yi}^2 + h_{zi}^2}$ is the linear momentum module, $p = a(1\text{-e}^2)$ is the satellite orbit focal parameter, i is the satellite orbit inclination angle, Ω

is the satellite orbit right ascension of ascending node angle (RAAN), $u$is the argument of latitude angle, ν is the satellite true anomaly angle, and ω is the satellite orbit argument of perigee angle (**Figure 27**).

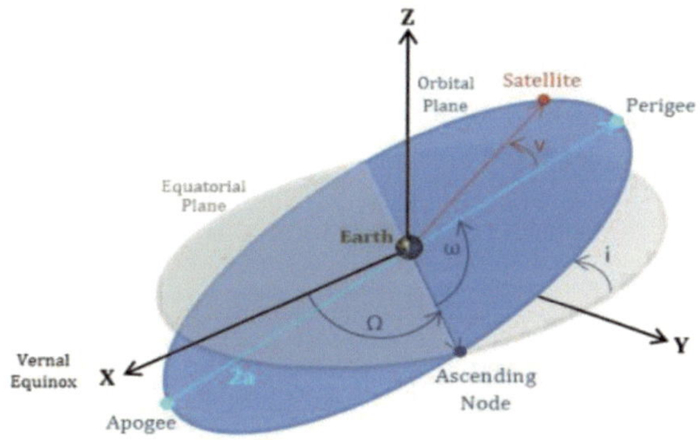

Figure 27.
Satellite orbit in the inertial ECI (XYZ) coordinate system.

3.2.3.2.2 Orbital thrusters control

If orbit maneuver is required, then it can be commanded by AODCS SW autonomously, or special control commands TLM (uploaded command tables) are sending to satellite AODCS, and in predetermined time they are executed activating at scheduled time for the calculated period (Δt) the orbital thrusters that provide for the required orbital correction/maneuver pulse$(F\Delta t)$.

3.2.3.3 ADCS SW

3.2.3.3.1 Satellite attitude and angular velocity estimation algorithms

This group of algorithms was presented above in 3.2.1.2 and can be used here.

For example, let us consider single-axis stabilized satellite that should keep one axis (e.g., Z) permanently pointed to the Sun as in **Figure 28**. Only two angles of the satellite deviation from this direction and their angular velocities are required to know to point and keep it in this direction. The satellite has two-axis Sun sensor that

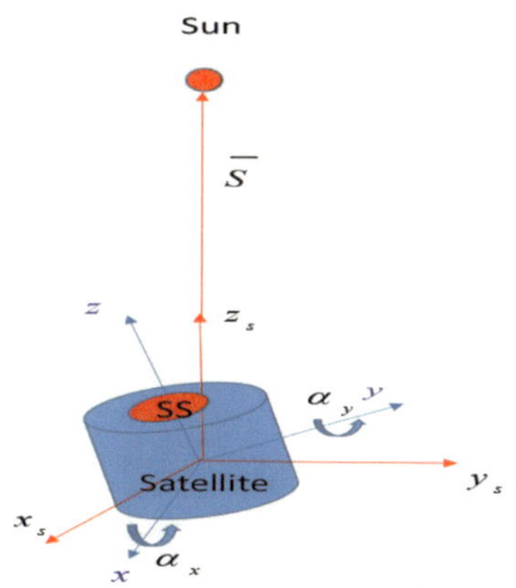

Figure 28.
Satellite pointed by the Z-axis to the sun.

can measure two angles α_x, α_y of satellite deviation from the sun direction. Its axes coincide with the satellite axes xyz. The axis z is the sensitivity axis that nominally should point into the Sun's direction (center of brightness), and xy is the focal plane. The Sun vector is referenced in the Sun frame as $S_r = [0\ 0\ \mathrm{S}]^T$ and is measured in the Sun sensor measured frame as $S_m = \left[S_x\ S_y\ S_z\right]^T$.

In **Figure 28**, xyz is the satellite body frame, $x_s y_s z_s$ is the Sun reference frame, SS is the two-axis Sun sensor and α_x and α_y are the turn angles of satellite x and y axis accordingly.

The following formula represents the mathematical transformation of the Sun vector from the reference into the body frame:

$$\mathbf{S_m} = \mathbf{C_{bs}} \mathbf{S_r} \tag{41}$$

where $\mathbf{S_r} = [0\ 0\ \mathrm{S}]^T$, $\mathbf{S_m} = \left[S_x\ S_y\ S_z\right]^T$, and $\mathbf{C_{bs}}$ is the DCM between the reference (Sun) and measured (satellite) frames. Let us consider that the order of rotation from the Sun to the satellite frame is 3-2-1 ($\alpha_z, \alpha_y, \alpha_x$); the DCM matrix $\mathbf{C_{bs}}$ is as follows [9]:

$$\mathbf{C_{bs}} = \begin{bmatrix} \cos\alpha_y \cos\alpha_z & \cos\alpha_y \sin\alpha_z & -\sin\alpha_y \\ -\cos\alpha_x \sin\alpha_z + \sin\alpha_x \sin\alpha_y \cos\alpha_z & \cos\alpha_x \cos\alpha_z + \sin\alpha_x \sin\alpha_y \sin\alpha_z & \sin\alpha_x \cos\alpha_y \\ \sin\alpha_x \sin\alpha_z + \cos\alpha_x \sin\alpha_y \cos\alpha_z & -\cos\alpha_x \cos\alpha_z + \cos\alpha_x \sin\alpha_y \sin\alpha_z & \cos\alpha_x \cos\alpha_y \end{bmatrix} \tag{42}$$

Then from (41), (42) can derive the following formulas:

$$\begin{cases} S_{xm} = -S \sin\alpha_y \\ S_{ym} = S \sin\alpha_x \cos\alpha_y \\ S_{zm} = S \cos\alpha_x \cos\alpha_y \end{cases} \tag{43}$$

From (43), desired angles and can be derived that can be used for satellite attitude control.

$$\begin{cases} \alpha_x = \tan^{-1}\dfrac{S_{ym}}{S_{zm}} \\ \alpha_y = -\sin^{-1}\dfrac{S_{xm}}{\sqrt{S_{ym}^2 + S_{zm}^2}} \end{cases} \tag{44}$$

3.2.3.3.2 Angular velocities (body rates)

Let us also assume that the satellite does not have angular velocity sensors RS and its angular velocities should be derived from the measured angles α_x and α_y. Simple low-frequency first-order differentiating fitters can be applied for this purpose. Laplace operator s-form (transfer functions) of these filters are presented below:'

$$\begin{cases} \hat{\dot{\alpha}}_x = \dfrac{s}{T_{fx}s + 1}\alpha_x \\ \hat{\dot{\alpha}}_y = \dfrac{s}{T_{fy}s + 1}\alpha_y \end{cases} \tag{45}$$

where T_{fx}, T_{fy} are filter time constants, typically, $T_f < (3 - 10)\,s$.

3.2.3.3.3 Satellite control algorithms

If not the optimization criterion to characterize the control quality [13] is required, then conventional negative feedback closed control loop with linear PID (proportional, integral, and damping) control law [9] that provides a good performance for many practical satellite control applications can be used to satisfy the requirements. They are typical for any automatic control system requirements: such as transfer process decay time and overshooting, residual static error caused by the permanent external disturbance, etc. Today, attitude control system performance can be verified mainly on ground with simulation. If we try to evaluate it in flight, then only onboard attitude sensors TLM data can be used for postprocessing, and it should be taken into account that mainly sensors that detect high-frequency noise (perceived errors) will be observable and low-frequency components (sensor biases) are compensated in the closed control attitude stabilization loop. Simple example of single-axis satellite attitude stabilization control loop is presented below. It is a simplified linear model; however, it presents the stabilization principle and essential features. Let us assume that a simple, positional, and damping control law is used to stabilize satellite axis Z in Sun direction $\overline{S}$ as in the example C1 for attitude and angular rate determination above. Let us assume that only the one-axis control channel X is considered; small angle α_x $(\tan\alpha_x \approx \alpha_x)$ is measured with SS (44), and its rate is derived with linear differentiating filter (45). Then requested PD control torque can be presented by the following formula:

$$T_{cx} = -k_{px}\alpha_{xm} - k_{dx}\frac{s}{T_{fx}+1}\alpha_{xm} \tag{46}$$

where k_{px} is the position control coefficient, k_{dx} is the damping control coefficient, $s = \frac{d}{dt}$ is the Laplace operator, T_{fx} is the differentiating filter time constant, $\alpha_{xm} = \alpha_x + \Delta\alpha_x$ is the measured angle α_x,α_x is the true value and $\Delta\alpha_x$ is the measured error. Let us assume that this torque is generated by only one MTR Y. Eq. (27) is determined as:

$$T_{cx} = B_z M_y = B_z k_{TRy} u_y \tag{47}$$

where $B_z = const$ is the local Z component of Earth magnetic induction vector, $M_y = k_{TRy}u_y$ is the Y MTR magnetic moment, k_{TRy} is the Y MTR control gate and i_y is the control voltage applied to Y MTR winding. Then as it follows from (46), (47) requested from the AODCS OBC control voltage to the winding of Y MTR is:

$$u_y = -K_{px}\alpha_{xm} - K_{dx}\frac{s}{T_{fx}+1}\alpha_{xm} \tag{48}$$

where $K_{px} = \frac{k_{px}}{B_z k_{TRy}}$ and $K_{dx} = \frac{k_{dx}}{B_z k_{TRy}}$ are position and damping magnetic control coefficients.

Let us take a ball-shaped satellite with the inertia matrix as follows: $J = \begin{bmatrix} J_x & 0 & 0 \\ 0 & J_y & 0 \\ 0 & 0 & J_z \end{bmatrix}$ where $J_x = J_y = J_z J_x = J_y = J_z$. Then in inertial spatial, its linear angular dynamical equations for the axis X can be approximately written as follows:

$$J_x s^2 \alpha_x = T_{c_x} + T_{d_x} \tag{49}$$

where $s = \frac{d}{dt}$ is the Laplace operator, T_{c_x} is the control torque, and T_{d_x} is the disturbing external torque (satellite residual and induction magnetism torque, atmosphere drug torque, solar pressure torque). Then substituting in Eq. (49) Eqs. (47) and (48), we can rewrite it as follows:

$$J_x s^2 \alpha_x = -k_{px}\alpha_{xm} - k_{dx}\frac{s}{T_{fx}s+1}\alpha_{xm} + T_{dx} \tag{50}$$

Let us divide all terms in Eq. (50) by the coefficient k_p and substitute α_{xm} value, then it can be represented in the following form:

$$\left(T_x^2 s^2 + 2d_x T_x \frac{s}{T_{fx}s+1} + 1\right)\alpha_x = \frac{1}{k_{px}} T_{dx} - \Delta\alpha_x \tag{51}$$

where $T_x = \sqrt{\frac{J_x}{k_{px}}}$ is the X control channel time constant and $d_x = \frac{k_{dx}}{2\sqrt{k_{px}J_x}}$ is the Xcontrol channel-specific damping coefficient. Eq. (51) is a third-order linear differential equation and could be analytically analyzed. In particular its stability can be analyzed with algebraic Hurwitz criterion [13]. However, more simple and general results can be obtained with the following approximate consideration. If $T_{fx} << T_x$, then by the filter time constant, T_{fx} can be neglected, and (51) can approximately be considered as a standard second-order control unit, presented by the second-order linear time invariant (LTI) differential equation and rewritten as follows:

$$\left(T_x^2 s^2 + 2d_x T_x s + 1\right)\alpha_x \simeq \frac{1}{k_{px}} T_{dx} - \Delta\alpha_x \tag{52}$$

As it follows from Eq. (52), steady-state error in attitude stabilization can be calculated with the formula:

$$\left(T_x^2 s^2 + 2d_x T_x s + 1\right)\alpha_x \simeq \frac{1}{k_{px}} T_{dx} - \Delta\alpha_x \tag{53}$$

where $T_{dx0} = const$ and $\Delta\alpha_{x0} = const \Delta\alpha_{x0} = const$ (ATT sensor bias).

For Eq. (52), *the optimal* damping coefficient is $d_x = \frac{\sqrt{2}}{2} = 0.707$ [13].

Numerical example

Let us evaluate satellite time constant T_x. Let us assume that for the LEO satellite magnetic field induction vector $\overline{B}$ has the following value of the projection on the Sun direction.

and MTR has the following parameters: maximal magnetic moment $M_{y\,\max} = 35Am^2$, maximal control current $I_{y\,\max} = 100$ mA, winding resistance $R = 280$ Ohm; and maximal control voltage $u_{y\max} = R \cdot I_{y\,\max} = 28$ V. MTR gate is $k_{TRy} = \frac{M_{y\,\max}}{u_{y\,\max}} = \frac{35Am^2}{28V} = 1.25Am^2/V$. Then maximal available magnetic torque is $T_{cx\,\max} = 10^{-3}Nm$. If maximal linear zone for this control channel is $\alpha_{y\,\max} = \frac{\pi}{2}$ rad, then the position control coefficient k_{px} can be calculated with the following formula:

$$k_{px} = \frac{T_{cx\,\max}}{\alpha_{\max}} \tag{54}$$

For the data above, it has the value of $k_{px} = \frac{0.001}{3.14 \cdot 0.5} = 6.366 \cdot 10^{-4}$ Nm/rad. Let us consider example of the first Soviet satellite "Sputnik" (SS-1) that had the mass $m = 83.6$ kg and the radius$R = 0.29$ m and take the assumption that its mass was homogeneously distributed within its spherical volume of $V = \frac{4}{3}\pi R^3 = 0.1\ \text{m}^3$, and then its inertia ($J = J_x = J_y = J_z$) can be calculated with the following formula:

$$J = \frac{2}{5} m R^2 \tag{55}$$

as for a homogeneous sphere. Substituting into Eq. (55) the data above, we can calculate that for SS-1 $J = 2.82\ \text{kgm}^2$, then its time constant with the considered MTR control might be:

$$T_x = \sqrt{\frac{J_x}{k_{px}}} = \sqrt{\frac{2.82 kgm^2}{6.366 \cdot 10^{-4} kgm^2 s^{-2}}} = 66.56\ \text{s}.$$

Now damping coefficient can be calculated with the following formula:

$$k_{dx} = 2 d_x T_x k_{px} \tag{56}$$

It has the following value: $k_{dx} = 2 \cdot 0.707 \cdot 66.56 \cdot 6.366 \cdot 10^{-4}\ \text{s} \cdot \text{Nm/rad} = 0.06\ \text{Nms/rd}$.

Finally, K_{px} and K_{dx} can be calculated. They are as follows: $K_{px} - \frac{k_{px}}{B_z k_{TRy}} = \frac{6.366 \cdot 10^{-4} Nm/rad}{2.86 \cdot 10^{-5} \cdot 1.25 T \cdot A \cdot m^2/V} = 17.8\ V/rad$ and $K_{dx} = \frac{k_{dx}}{B_z k_{TRy}} = \frac{0.06 Nms/rad}{2.86 \cdot 10^{-5} \cdot 1.25 T \cdot A \cdot m^2/V} = 1.678 \cdot 10^3 sV/rad$.

When T_x is determined, then the time constant for the differencing filter T_{fx} can be chosen from the condition that $T_{fx} < < T_x$. In our example $T_x = 66.56$ s, let us take that $T_{fx} = 5$ s.

Simulation

Eqs. (51) and (52) were simulated using Simulink (see **Figure 29**).

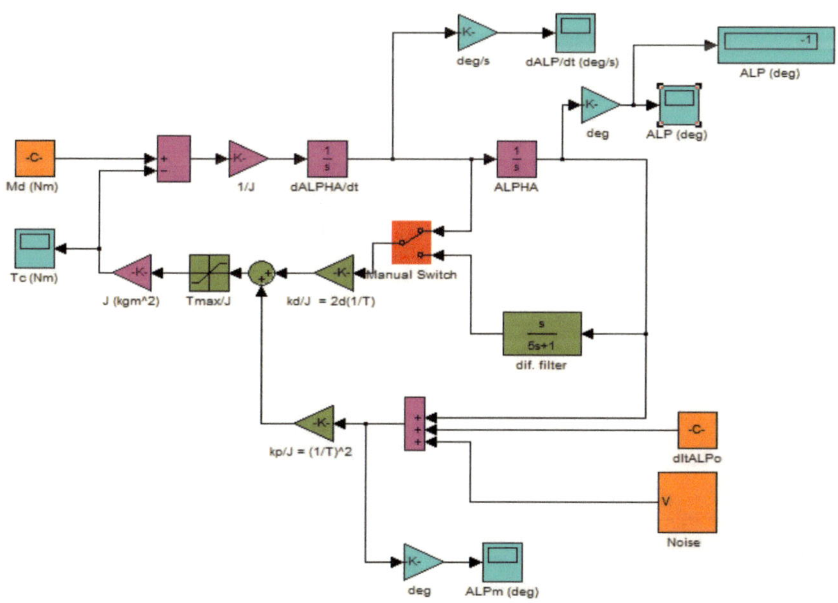

Figure 29.
Satellite single-axis attitude control Simulink block scheme.

Blocks in the pink color present the satellite model, the dark green color is for control law blocks, the cyan blocks are registration oscilloscopes, and the display and the orange color are the disturbances. The red manual switch allows to implement the differentiating filter, transforming the scheme from the approximation (52) to the accurate presentation (51). Disturbing external torque M_d is constant; attitude sensor error is represented by the constant value ALP_0 and limited range white noise V that has spectral density $S_V = 2\sigma_V^2 T_v$ (σ_v is the standard deviation (SD), T_v is the correlation time). The results of the simulation (**Figure 29**) with and without the differentiating filter (with the assumption that $\dot{\alpha}_x\dot{\alpha}_x$ is directly measured without any errors) are presented below in **Figures 30–34** (left A, (52), without the filter; right B, (51), with the filter). The numerical data for the simulation are as follows:

$$J = 2.82\ \text{kgm}^2, T_x = \sqrt{\frac{J_x}{k_{px}}} = 66.56\ \text{s}, d_x = \frac{k_{dx}}{2T_x k_{px}} = 0.707,$$

$$k_{px} = 6.366 \cdot 10^{-4}\ \text{Nm/rad}, k_{dx} = 0.06\ \text{Nm} \cdot \text{s/rad}, T_{fx} = 5\ \text{s},$$
$$M_d = 10^{-5}\text{Nm}, \Delta\alpha_{x0} = dltALP_0 = 0.1^{\circ}, \Delta\alpha_{x0} = dltALP_0 = 0.1^{\circ}$$

Simulation of ACS (**Figure 29**) is presented in **Figures 30–34**. Units: vertical axis (deg), horizontal axis: (s).

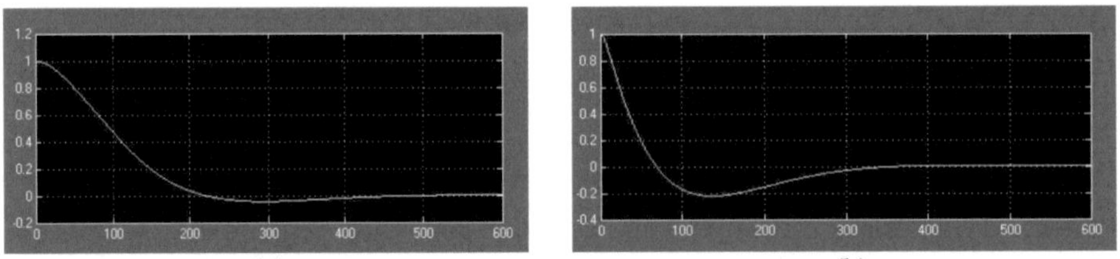

(a) (b)

Figure 30.
Response to initial deviation angle $\alpha_{x0} = 1°$. (a) without dif. filter and (b) with dif. filter.

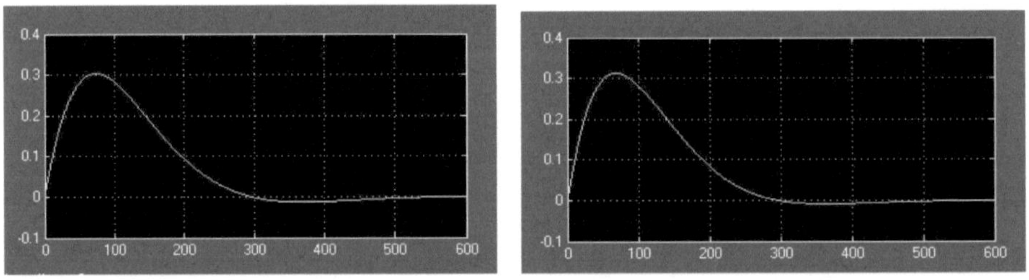

Figure 31.
Response to initial angular velocity $\dot{\alpha}_{x0} = 0.01\text{deg/s}$.

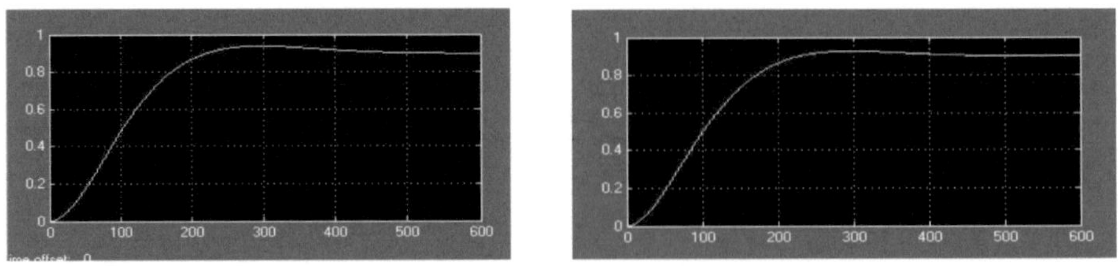

Figure 32.
Response to external disturbance torque $T_d = 10^{-5}\text{Nm}$. Static error $\alpha_x^* = 0.898°$.

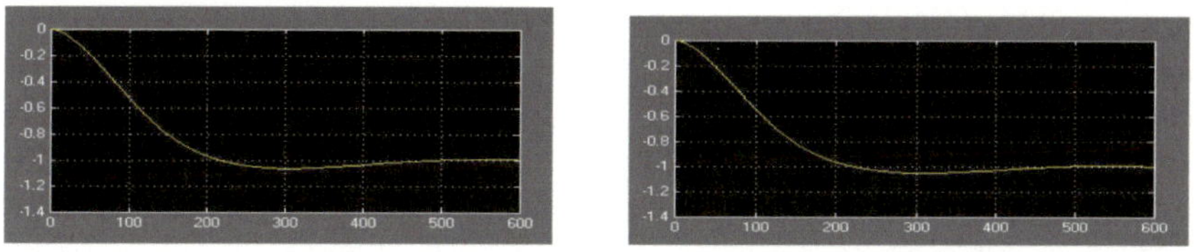

Figure 33.
Response to attitude sensor bias $\Delta_{xo} = 1^{\circ}$ *plus white noise* $\sigma_V = 0.1^{\circ}, T_V = 1$ s. *Satellite attitude stabilization errors, ALP.*

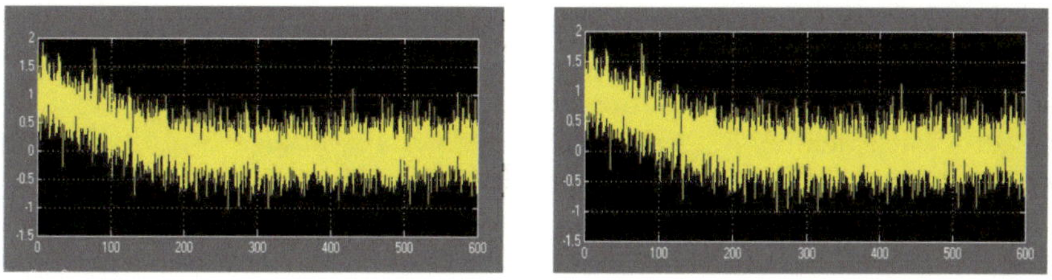

Figure 34.
Satellite attitude measured errors ALPm.

Decay time: $\tau = 195s, \alpha_x(3\tau) = 5\%\alpha_{x0}(\text{deg})$, (s) $\tau = 275s$.

As it can be seen, measured noise is filtered effectively in the control loop, and stabilization error is equal to the sensor bias with opposite sign.

In **Figure** 34, we can see that the measured (perceived) errors that TLM data provide to ground after the decay time do not present sensor bias and present only measured noise. It is because satellite stabilization error with opposite sign compensates the bias. In general, it can also be seen that the simulation of the approximate second-order model (52) is very close to the accurate model (51). Hence, at least for the analytical representation, (52) can be successfully used.

4. Conclusion

Part I of this chapter presents an overview of practical satellite control system, satellite guidance, navigation and control equipment. The work presented here is based on the author's point of view of integration of this GN&C equipment in the integrated AODCS system (satellite GN&C Spacetronics System). Main work principles, architecture, and components of the satellite control system were briefly highlighted.

The chapter can serve to a wide pool of space system specialists as an introduction to satellite control system development.

Acknowledgements

The author wishes to express his sincere gratitude to the Canadian Space Agency, where he had the opportunity to learn and possess the knowledge and experience related to the writing of this chapter. As well, he is very thankful to many of his colleagues from Canadian Magellan Aerospace Company (Bristol Aerospace Division) with whom he discussed and analyzed satellite AODCS design projects and issues that helped him to work out the system analysis and its principal concepts presented in this chapter. Additionally, he cannot forget that his experience and background in Aerospace Technology were also accumulated from the former USSR (Moscow Aviation Institute, Moscow Aviapribor Corporation, Moscow Experimental Design Bureau Mars, Institute in Problems in Mechanic of RAN) and Israel (IAI, Lahav Division and Tashan Engineering Center), where he could observe and learn from diverse and wealthy engineering and scientific schools led by great scientists and designers such as Prof. BA. Riabov, Prof. V.P. Seleznev, V.A. Yakovlev, G. I. Chesnokov, V.V. Smirnov, Dr. A. Syrov, Acad. F. Chernousko, A. Sadot and Dr. I. Soroka.

This chapter was written as a solo author since his friend and regular coauthor Prof. George Vukovich from York University of Toronto passed away 2 years ago. For many years, Prof. Vukovich served as Director of his Department of Spacecraft Engineering in CSA. He will always keep good memories of Prof. Vukovich who helped and encouraged him continue his scientific and engineering work.

The author also acknowledges the copyrights of all publishers of the illustrations that were extracted from the open sources in the Internet.

Note

Dedicated to Prof. G. Vukovich.

Author details

Yuri V. Kim
David Florida Laboratory, Canadian Space Agency (CSA), St. Hubert, Ottawa, Canada

*Address all correspondence to: yurikim@hotmail.ca

References

[1] Eickhoff J. On Board Computers, on Board Software and Satellite Operation. Berlin: Springer-Verlag; 2012

[2] Space Systems, Design, Qualification and Acceptance Tests of Small Spacecraft and Units. International Standard, ISO 19683; 2017

[3] Space Engineering, Satellite Attitude and Orbital Control System (AOCS) Requirements, ECSS-E-ST-60-30C; 2013

[4] Development of First Artificial Satellites (Rus.). Available from: https://www.kik-sssr.ru/Hist_1_PS-1_books.htm [Accessed: 15 January 2020]

[5] Available from: https://www.space.com 17563-sputnik. Sputnik: The Space Race's Opening Shot [Accessed: 17 January 2020]

[6] Chaturvedi A. Do You Know How Many Satellites Are Currently Orbiting Around the Earth? Available from: https://www.geospatialworld.net/blogs/do-you-know-how-many-satellites-earth/ [Accessed: 16 January 2020]

[7] About the David Florida Laboratory-Canada.ca, Available from: https://www.asc-csa.gc.ca/eng/laboratories-and-warehouse/david-florida/about.asp [Accessed: 17 January 2020]

[8] Delchamos T, Jonason G, Swift R. The Spacecraft Test and Evaluation Program, Telestar I, NASA SP-32. Washington, DC; 1963. p. 1007

[9] Sidi M. Spacecraft Dynamics and Control. A Practical Engineering Approach. Cambridge: Cambridge University Press; 1997

[10] Wertz JR, editor. Spacecraft Attitude Determination and Control. Dordrech: Kluwer Academic Publishers; 1978

[11] You Z. Space Microsystems and Micro/Nanosatellites, National Defence Industry Press. NY: Elsevier Inc; 2018

[12] Lee J, de Rouiter A, Ng A, Lambert C, Kim Y, Yoshihara K. Attitude Determination and Control Subsystem of JC2Sat-FF Mission. In: Proceedings of 12th International Aerospace Conference ISCOPS in Montreal, July 2010, AAS 10-427; 2010

[13] Bryson AE, Yu-Chi Ho J. Applied Optimal Control. Levittown, PA: Taylor & Francis; 1975. pp. 364-373, 369, 457

[14] Fourati H, Belkhiat D. Multisensor Attitude Estimation. Fundamental Concepts and Applications. NW: CRC Press, Taylor & Francis Group; 2017

[15] Kim YV. Kalman filter decomposition in the time domain using observability index, IFAC, Seoul, July 6–11, 2008. In: International Conference Proceedings. Vol. 2. NY: Curran Associates Inc.; 2009. pp. 625-630

[16] Seleznev VP. Navigation Devises, Rus. Moscow: Mashinistroenye; 1974

[17] Chatfield A. Fundamentals of High Accuracy Inertial Navigation. Reston, VA: AIAA; 1974

[18] Titterton D, Weston J. Strapdown Inertial Navigation Technology. Reston, VA: The Institute of Engineering and Technology; 1996

[19] Hofmann-Welenghof B, Lichtenegger H, Collins J. GPS Theory and Practice. New York: Springer-Verlag Wien; 2001

[20] Shuster M, Oh S. Three-axis attitude determination from vector observations. Journal of Guidance & Control. 1981; **4**(1):70-77

[21] Armenise M, Ciminelli C, Dell'Olio F, Passaro V. Advances in Gyroscope Technologies. Berlin: Springer-Verlag; 2010

[22] Arnold R, Maunder L. Gyrodynamics and its Engineering Applications. London: Academic Press; 1961

[23] Meshchersky IV. Dynamics of a point of variable mass. In: Work on the Mechanics of Bodies of Variable Mass. 2nd ed. GITTL, Moscow; 1952. pp. 37-188, 280

[24] A Man- and an Equation, ESA. Available from: http://blogs.esa.int/rocketscience/2012/10/14/a-man-and-an-equation/ [Accessed: 19 January 2020]

[25] How to Slow Down Satellite Rotation? D-dot Algorithm, ESE 3 Sat Project. Available from: http://www.ece3sat.com/blog/2018-01-25-how-to-slow-down-rotation-the-bdot-algorithm/

[26] Space Engineering-Software, ECSS-E-ST-40C, Noordwijk; 2019

[27] Vukovich G, Kim Y. Satellite orbit deacay due to atmospheric drag. International Journal of Space Science and Engineering. Inderscience Publishers; 2019;**5**(2). Availabe from: www.inderscience.com/jhome.php?/jcode=ijspacese. [Acessed: 4 June 2020]

Chapter 4

Game Theoretic Training Enabled Deep Learning Solutions for Rapid Discovery of Satellite Behaviors

Dan Shen, Carolyn Sheaff, Genshe Chen, Jingyang Lu, Mengqing Guo, Erik Blasch and Khanh Pham

Abstract

The chapter presents a game theoretic training model enabling a deep learning solution for rapid discovery of satellite behaviors from collected sensor data. The solution has two parts, namely, Part 1 and Part 2. Part 1 is a PE game model that enables data augmentation method, and Part 2 uses convolutional neural networks (CNNs) for satellite behavior classification. The sensor data are propagated with the various maneuver strategies from the proposed space game models. Under the PE game theoretic framework, various satellite behaviors are simulated to generate synthetic datasets with labels for the training to detect space object behaviors. To evaluate the performance of the proposed PE model, a CNN model is designed and implemented for satellite behavior classification. Python 3 and TensorFlow are used in this implementation. The simulation results show that the trained machine learning model can efficiently and correctly classify the satellite behaviors up to 99.8%.

Keywords: space situational awareness, satellite characterization, sensor models, simulated training data, training performance, general-sum games, CNN

1. Introduction

Since space has already been fully utilized, society has become increasingly dependent on space advantages in various industrial, civil, and commercial applications. This dependence brings an essential vulnerability, especially shortage of continuous situational awareness of the space environment to ensure freedom of movement. Due to the fact that information from space is crucial for key decision-making, such as urban, agricultural, and responsive planning, space is regarded as a significant frontier. In addition to real-time and hidden information constrains, the existence of space object density significantly produces the complexity of the space situational awareness (SSA). Understanding the position of space objects from low-level information fusion can support high-level information fusion SSA missions of sensor, user, and task refinement [1]. In order to implement SSA accurately, it is possible to coordinate the evaluation of residential space objects (RSO) through user-defined operation pictures (UDOP) [2].

Space control and SSA are required for space prevalence, which depend on fast and precise space object behavioral discovery. Developing a theoretical approach for fast discovery of variation of satellite behaviors is the main task of this book chapter. The machine learning methods with novel neural networks are proposed in this case. However, there are numerous challenges for constructing the tools because of the following reasons: (i) partially observable movements, (ii) resident space objects (RSOs), (iii) uncertainties modeling and propagation, (iv) real-time response, and (v) computationally intractable algorithms.

Space access investigation and mission trade-off considers are imperative for the victory of space-borne operations. The tracking algorithms of space object can be measured depending on collecting data to track satellites, debris, and natural phenomena (such as comets, asteroids, and solar flares). Tacking is related with sensor administration, which can point sensors to observation points to decide the circumstance and to aware threatens. SSA improvements consist of models (such as orbital mechanics), measurements, computational software (such as tracking), and application-based system coordination (such as situations). For instance, *game-theory* methods for SSA can be utilized for pursuit-evasion analysis [3].

This book chapter creates and establishes game theoretic training enabled deep learning (GTEL) methodologies for fast discovery of the behaviors from the satellites. GTEL is an adaptive feedback adversarial theoretic method, which acquires data from sensors related to the relationship between resident space objects (RSOs) of interest and sensing assets from ground and space (GSAs). Thus, a game theory is modeled instead of a control problem for this circumstance. Game reasoning uses data-level fusion, random modeling/propagation, on the other hand, RSO detection/tracking predicting the future RSOs-GSAs relationships. The adversarial engine also supports optional space pattern dictionary/semantic rules for adaptive transition in the Markov game. In the event that no existing pattern dictionary is accessible, GTEL will construct an initial pattern and modify it during the game inference. The output of GTEL inference consists of two parts of control methods: (i) measurements processing and (ii) RSOs localization. The two parts establish a *game-equilibrium*, one of which is sensing asset management, the other of which is the estimation of RSO behaviors.

The chapter is organized as the following. Section 2 introduces the comprehensive system design for our methods. Section 3 presents the Markov game theory with satellite maneuvering. Section 4 proposes the details of our machine learning model for space behavior detection. And the numerical results and analysis are displayed in Section 5. At the end, Section 6 draws the conclusion of this chapter.

2. Overall system architecture

The proposed methodology of GTEL is shown as **Figure 1**. The core piece of the method for detecting ***unknown*** patterns of space objects is Markov Game Engine. Due to the patterns are obscure, there is no preparing training data accessible. Therefore, the Markov Game Engine makes use of zero-shot learning [4] and transfer learning to unsupervised classify unavailable target domain data by training available data from source domain (i.e. simulate data). The knowledge adaptation or domain transfer is performed through manifold learning to share intermediate semantic embeddings (such as attributes) between labeled and unlabeled data. On one hand, the manifold learning can reduce the dimension of sensing data (such as azimuth angle, elevation angle, range, and range rate). On the other hand, the manifold learning can also mitigate the difficulties of object tracking and detection with fast space object behaviors' detection. Moreover, in order to solve the ***uncertainties*** in this task, the $\mathbb{R}^5 \times \mathbb{S}$ coordinate system is used with filtering technology on *optimal transport (OT)*. The essential components of our methodology are shown as the following:

1. *Markov game* – The space conflicting situation is inferred by the Markov game structure [5, 6], in which system state is represented by distributions rather than deterministic values.

2. *Uncertainty modeling and propagation* – The uncertainties are expressed by the production of Von-Mises and independent Gaussian distribution for both measurement interference and original state uncertainty that are identified on cylindrical manifold $\mathbb{R}^5 \times \mathbb{S}$ [7]. Convergence is not guaranteed with relaxed synchronization for uncertainty spread of information.

3. *Optimal transport based tracking of space objects in cylindrical manifolds* – Compared with ensemble Kalman filter (EnKF) methods for tracking space objects, optimal transport (OT) [8, 9] is much more precise and robust. Furthermore, it is way broader than the algorithms with uncertainties assumption in $\mathbb{R}\,n$, because OT is not related to the distribution (since OT is focused on a transformation instead of importance sampling).

4. *Course of actions for behavior modeling* – The behavior of RSO is specified a course of action (CoA), which in turn dictates to what the RSO may appear to perform or is doing within subsequent a couple of steps.

5. *Manifold learning for data level sensor fusion* – The raw sensor data is generally high-dimensional. Considering that the measurement data is reflections of witnessed satellites (only several parameters will determine their states), so it can be reasonably assumed that the inherent dimension of the measurement domain is low. The communication bandwidth can be saved with decreasing the number of dimensions by using Manifold learning algorithms [10].

The simulated positions of satellite are used to generate sensor measurements, and the results will be utilized to track several space objects to complete the

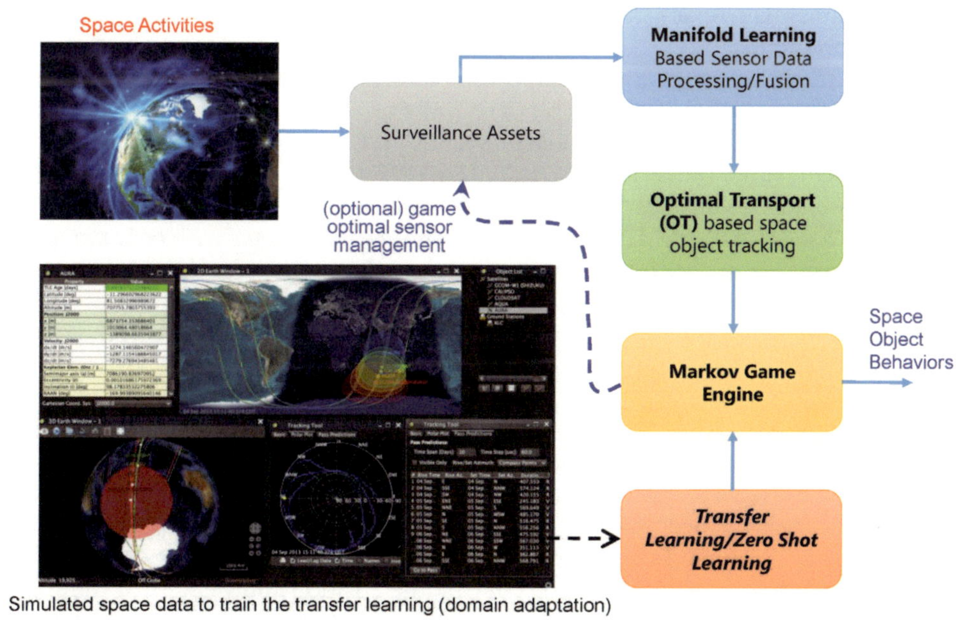

Figure 1.
GTEL system architecture.

estimation of the position. Afterwards, the estimations of the orbits are used for collision alert and maneuver detection of space objects. The satellite maneuvers are going to be interpreted as platform commands to perform course of actions and space object movements.

Figure 1 shows an adaptive feedback approach with the game theory enabled. It utilizes sensors to obtain information about the relationships between the ground/space sensing assets and the RSOs of interest [11]. RSOs and GSAs determine the relations. Thus, rather than saying it is a control problem, it is a dynamic game. Data-level fusion, game reasoning, RSO detection/tracking, and uncertainty modeling/propagation are combined to predict the RSO-SA relationships in the future. Optional space behavior dictionary/semantic rule for adaptive transition matrices in our Markov game is also supported by our game engine. In addition, the game solution is an equilibrium, which is controlled by both space sensing asset management and the (estimated) RSO behaviors.

3. Markov game in space situational awareness

Lloyd Shapley has invented a concept of stochastic game [6], which is a dynamic game played by one or several players focusing on probabilistic transitions. There are several stages for this game. At first, the game is set in one state. Then, the participated players should select an action individually. Based on the current state and the actions players chosen, each player will receive a reward. Therefore, after the chosen from each player, the game comes to a new random state, where the previous state and previous actions chosen by the players determine a distribution of the new random state. Afterwards, the above action will be repeated again for the new state. After finite or infinite number of stages of playing, the total reward for each player is obtained using the discounted sums of each stage reward or the averages of every stage rewards. In this way, each player gets a reward and the reward is compared with each other. The aforementioned sequences are the procedures for the stochastic game, which can be generalized by Markov decision processes (MDP) with repeated games. Our space situational awareness (SSA) would utilize this game tool for intent prediction [12].

The Markov game engine extracts specific information from each event as the following: (*i*) a finite set of players N, (*ii*) a finite or infinite set of states, S, (*iii*) a finite set of accessible actions for each player in N set, D^i (the overall action space is $D = \times_{i \in N} D^i$), (*iv*) a transition rule q: $S \times D \rightarrow \Delta(S)$, (where $\Delta(S)$ is the space of all probability distributions over S), and (*v*) a reward function r: $S \times D \rightarrow R^N$.

Figure 2 shows a visual description of the simple game states with only two players, who have only two options of actions for each player. The arrows in the graph indicates the probable transitions from one state to the other state. The states with red color means that only player 1 changes the approach. On the contrary, the states with blue color indicates only player 2 changes the approach. The state with green color indicates both players change the approach.

The GTEL solution uses a two-player Markov game to investigate the sensor management for tracking space objects. Whether deliberate or unintentional, some of space objects may cause confusion to observers (sensors) when the orbital maneuvers are performed. In general, spatial object tracking can be assumed as an optimal control problem (one side optimization) or a game problem (two side optimization). For the settings of optimal control, the position and velocity of the space objects will be calculated (filtered) as the states dependent on the measurement from the sensor. However, this method ignores the possibility that the space objects may alter their orbits purposely with intelligence. It may cause difficulties for the satellite to track the space objects. Therefore, the Markov game method provides a solution for these difficulties. In this approach, on the one hand, the observed satellite will utilize the tracking and sensing model to destroy the tracking estimations by confusing the observer. On the other hand, the observer figures out ways to minimize the uncertainties of tracking, where the uncertainties are dependent on the entropy of tracking.

In this chapter, the information uncertainty of the GTEL pursuit-evasion (PE) game method [13] was exercised with a circumstance of two satellites, one of which is Geostationary Earth Orbit – GEO (observed satellite), the other of which is space based Low Earth Orbit-LEO satellite (observer satellite). **Figure 3** provides an illustration of a space based optical (SBO) sensor measurement model. The angle from the line from the object to the SBO and the line from the sun with object is defined as Bistatic Solar Angle, represented by θ. The smaller the angle is, the stronger the lighting conditions. Therefore, it causes difficulties for observations when the angle is large because of saturation of lighting. As shown in **Figure 4**, the scenarios of light have shown. In the graph, the blue line is for the LEO orbit, the green line

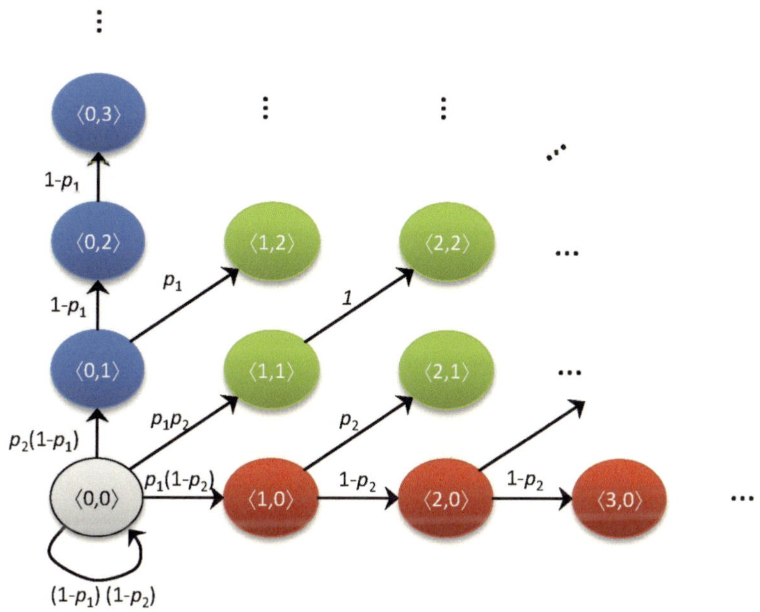

Figure 2.
A diagram of states in a Markov game.

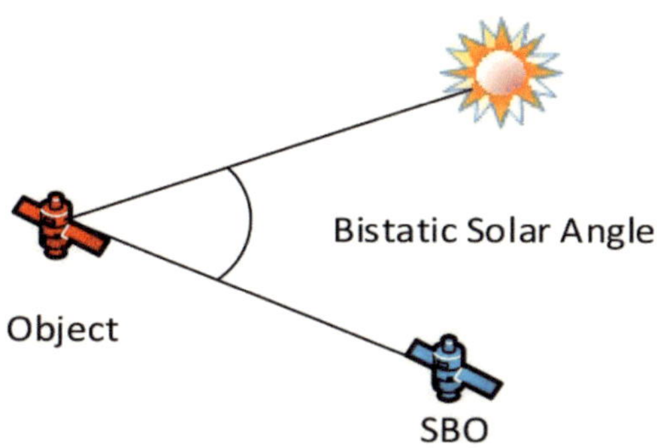

Figure 3.
SBO with a Bistatic solar angle.

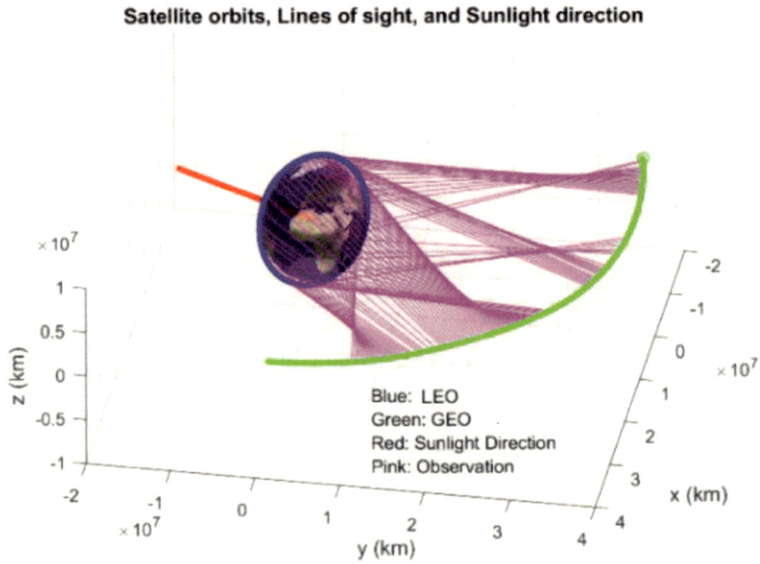

Figure 4.
LEO and GEO based on sensor management and maneuver strategies with game theory.

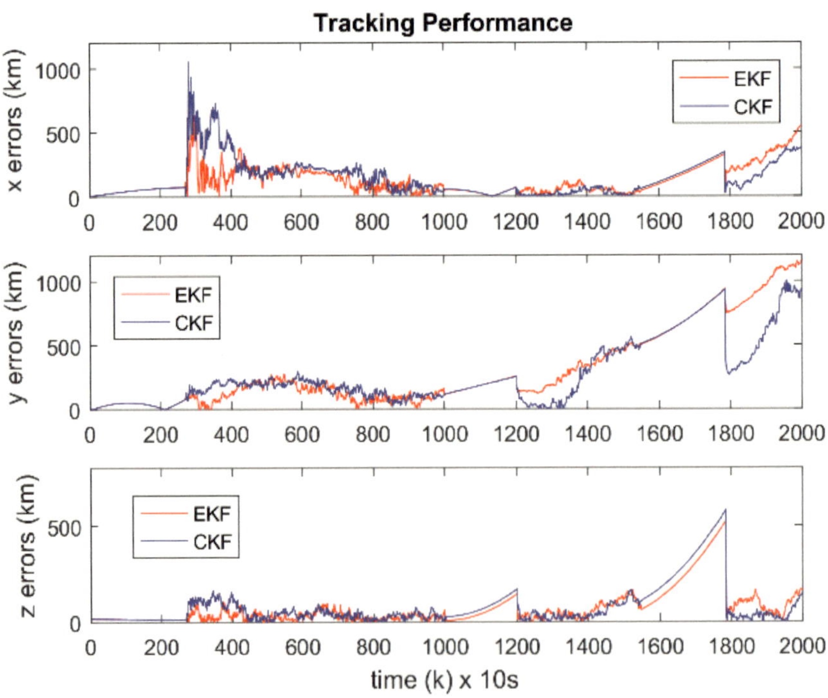

Figure 5.
The performance of tracking using theoretical Markov game strategies.

indicates the GEO orbit with maneuvers, the pink lines are the SBO sensor performed to track the GEO in order to lower the uncertainty, and the red line displays the sunlight from the sun to the earth.

The research scenario is shown in **Figure 4**, where the red line indicates the direction of sun light, green color is for the GEO orbit with maneuvers, blue color for LEO orbit, and pink lines indicate when the SBO sensor resource is used to track the GEO (use the sensor data to lower the uncertainty).

Figure 5 shows the results of tracking based on intermitted measurements. Both cubature KF (CKF) and extended Kalman Filter (EKF) trackers are shown. With the increase of the period without measurement due to the Earth blockage, the tracking errors increased as well. In addition, the tracking errors increased with the maneuver actions from the satellite. On the contrary, informational entropy decreased by sensor measures in this process.

Figure 6 top graph displayed the PE game control results, with α and β as angles of the maneuver thrust. The zoom-in view of the game optimal controls is shown

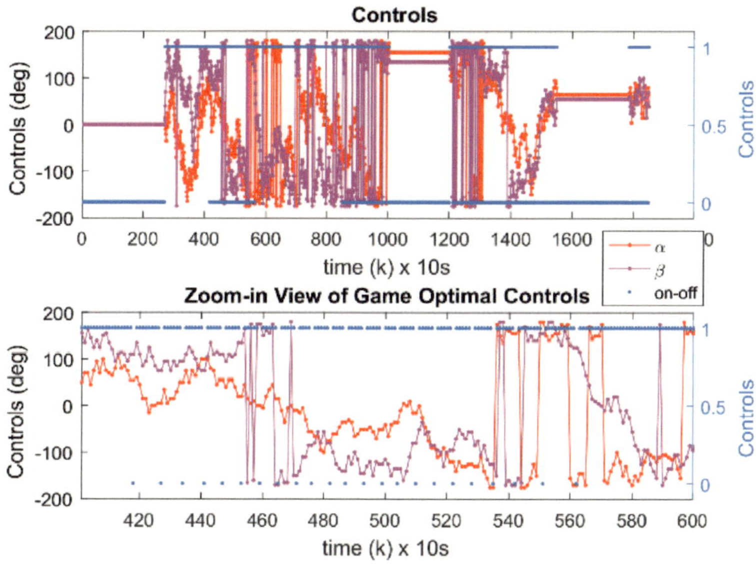

Figure 6.
The maneuver controls dependent on the PE game solution for the satellite direction.

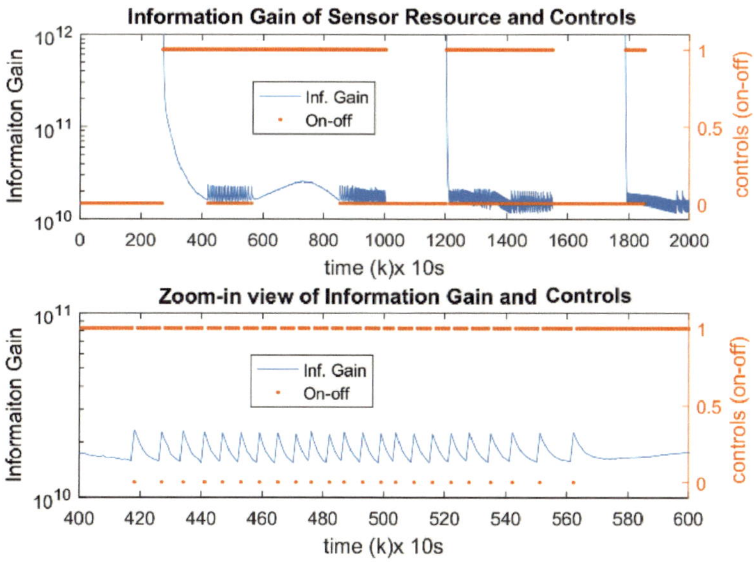

Figure 7.
Sensor's game theoretic on–off controls and associated information gains.

Figure 6 below, which exhibited that the observer satellite can keep the tracking uncertainty within a desired level while saving the resources of sensor. **Figure 7** shows the on–off sensor controls and the associated information gains. The results show that with the larger potential information gain, the sensor would use the resources to take measures (for the observer's on–off control, 0 means turning off and 1 indicates turning on).

4. Machine learning based RSO behavior pattern classification

4.1 The scheme of machine learning

The machine learning (ML) details will be described in this section. The main purpose of construction this machine learning scheme is to detect the behaviors of the resident space object (RSO) by fusing sensors data from multiple sources, including the velocity, orbital energy, angular momentum, and the position of RSO compared to the station. Similar with other machine learning model, this model is generally trained off-line by using generated data. Then the generalized weights will be deployed in the application using TensorFlow deployment. In addition, in order to improve the robustness of the trained system, we proposed a neural network scheme with the ability to train the newly-added unknown pattern online with only tiny modifications of the weights.

As shown in **Figure 8**, the RSO pattern classification architecture is displayed. It consists of two separate parts, one of which is offline part as *Modeling*, the other of which is online part as *Monitoring*, to detect the RSO behavior pattern.

As the matter of the offline part, named as *Modeling RSO Behavior Pattern*, our neural networks will be trained as a classifier by using the collected simulated data. The data specifically indicates the different behaviors of the RSO. As shown in the graph, some useful features are obtained from different sensors in the sessions of feature extraction. Subsequently, the extracted feature will be sent into the training model for neural network tuning and training to generate a classifier, which is used to identify the RSO behavior with fine grained size. On the contrary, the online part, *Monitoring RSO Behavior Pattern*, acquires RSO patterns in real-time to distinguish the abnormal behaviors. In this way, a warning message would be prompted if there are any abnormal behaviors detected.

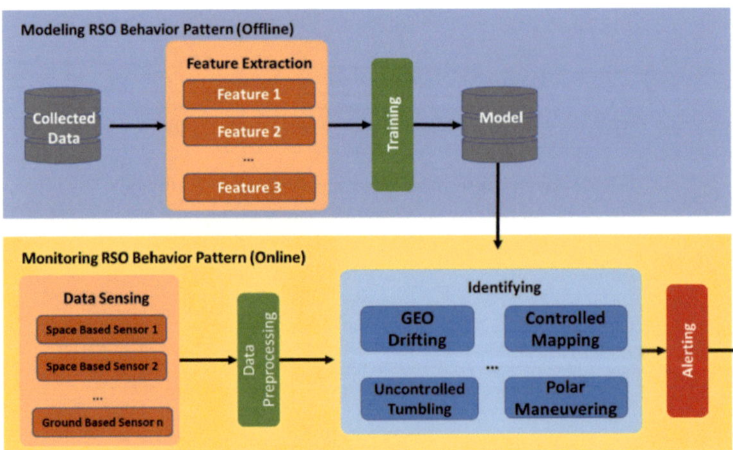

Figure 8.
The scheme of RSO behavior classification.

Additionally, the classifier generated by the machine learning methods has several properties as the following: (a) ability to fuse heterogeneous and complex input data, (b) scalability, (c) robustness with perturbations relative to the data, (d) high accuracy, and (e) explicitly. However, it is hard to fulfill all the requirements as shown above, therefore, some trade off would be considered for our model. Among different structures of neural networks, convolutional neural networks fit our case appropriately.

4.2 Framework of convolutional neural networks (CNNs)

The structures of Convolutional Neural networks (CNN) as well as Dense Neural Networks (DNN) are shown in **Figure 9**. The DNN neural networks consist of several hidden layers with hidden neurons. Every neuron in the subsequent layer is fully connected to the neurons in the previous hidden layer. The neurons contain the linear functions with activation function for each neuron, which is completely independent with the other neurons. Finally, after going through several layers, the input data is generated to the output data as the "output layer", which utilizes softmax activation function to produce classification probabilities.

However, the DNN (as top image shown in **Figure 9**) has several drawbacks, such as scaling difficulties for large images. For instance, for an image with $18 \times 18 \times 3$ dimensions, DNN for the first layers will have $18 \times 18 \times 3 = 972$ neurons. If the next DNN layer has 30 neurons, the weight parameters will be $972 \times 30 + 30 = 29190$. With a larger size of image, the weight parameters will increase a lot as well. Moreover, the 30 neurons may not have enough complexity to generalize the accuracy of our classification model.

On the other hand, the CNN (as bottom graph shown in **Figure 9**) can solve the aforementioned issues. A simple CNN [14] consists of several different filters as convolutional layer. In addition, the pooling layer can decrease the dimensions of the input data. Meanwhile, since the adjacent data in an image has similar values,

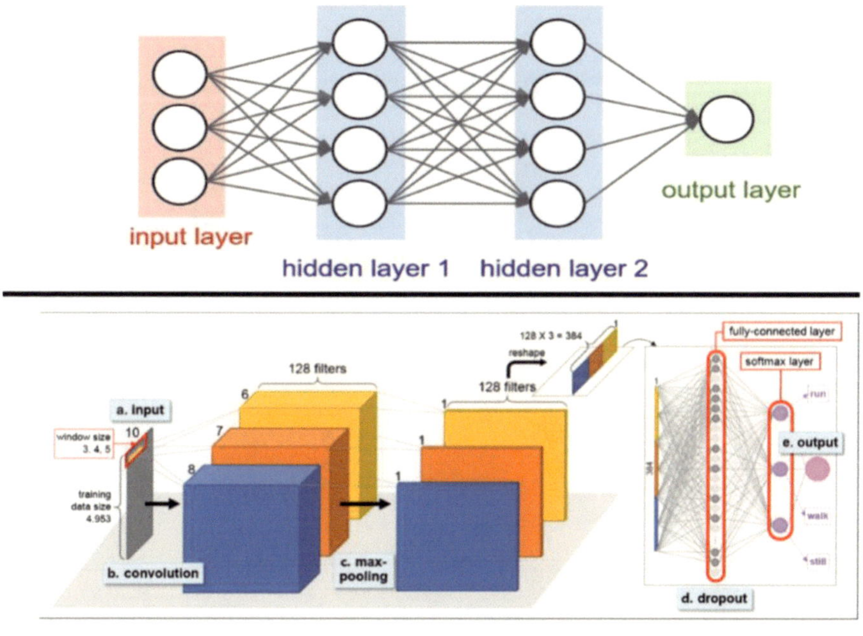

Figure 9.
The framework of the convolutional neural network.

the filter can extract the features reasonably with few weight parameters. Thus, using the CNN filters can increase the accuracy of our neural network classifier.

With the development of CNN, it successfully proceeds to be the best model structure for computer vision and image processing tasks. Moreover, CNN has also been used by Natural Language Processing (NLP) area. With small filters moving across the input dataset, the filters, with only small numbers of parameters, are re-utilized to recognize patterns for the large image. Therefore, with the similar classification abilities, CNN network is faster to train and predict compared with DNN. After several filters, the output part is flattened to bypass several dense layers to produce the softmax activation at the last layer with the sparse cross-entropy loss function to backpropagation for the behavior of RSO pattern prediction.

Our GTEL PE method utilizes the CNN architecture to classify the RSO pattern with observed data. Compared with the other conventional methods, the convolutional neural networks can process the RSO observation much faster. Therefore, we utilize the Python and TensorFlow with Keras [15] as the fundamentals for code write up. Although, each filter is computationally expensive to be trained, the overall CNN architecture is faster to be trained to provide similar accuracy classification.

A typical CNN structure is shown in **Figure 10**. After training our CNN-DNN with our training data, the test data (with 10–20%) will be employed to evaluate the generalization of our CNN-DNN model. Additionally, in order to solve the

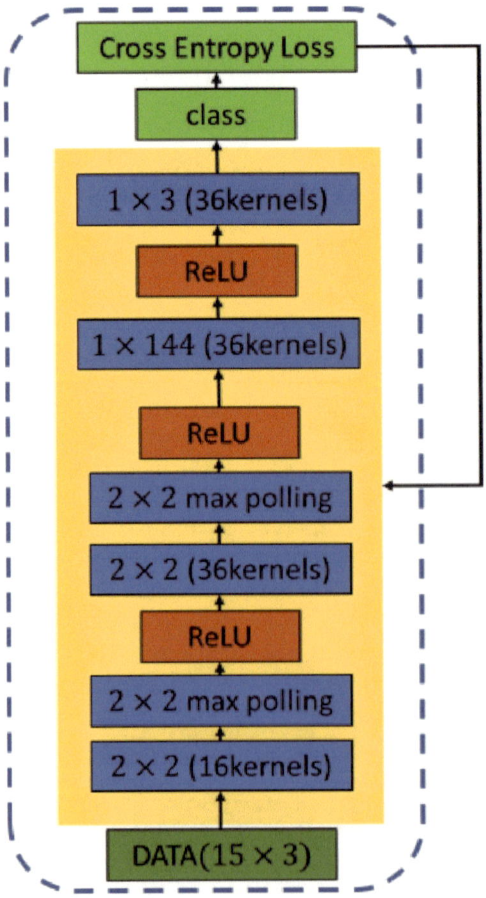

Figure 10.
CNN architecture for RSO behavior classification.

overfitting problem of our model, several dropout layers were added after each layer. This model outperforms other traditional methods with better accuracy and higher computationally efficiency.

The revised deep learning neural network structure for 143-sapce-behavior (we partition the pointing angles of α and β into cells with 15 degrees each and the total cell number is 143) is shown in **Figure 11**. At first, there are 3 different dimensions in our raw data, which are 3 parameters dimension, 15 times (each track consists of 15 observation measurements) interval dimension, and 1 number of channels dimension. Therefore, the convolutional neural network (CNN) comprises an initial 3*15*1 input dimension with 72,000 samples as the raw dataset. In order to distinguish the 143 different labels of satellite behaviors (15-degree difference for between each behavior), 128 "2 × 2" filters are utilized with the same padding for the first convolutional layer. Thus, there are 3*15*128 dimensions of output

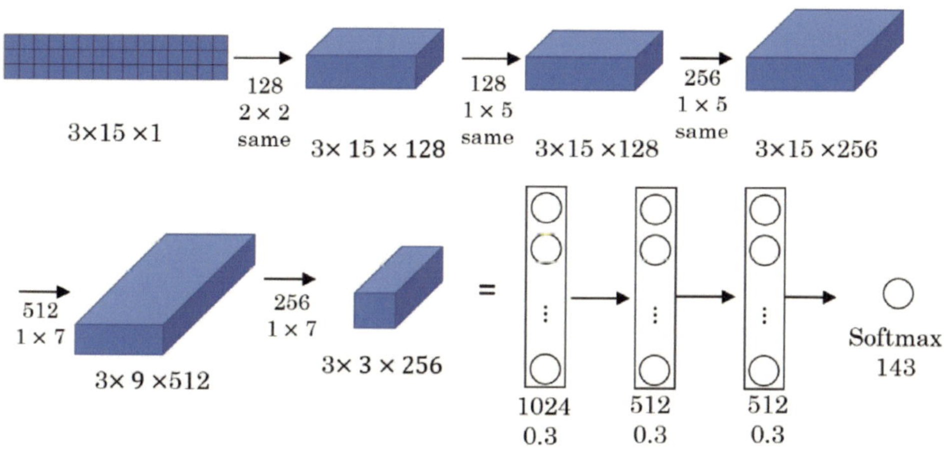

Figure 11.
Revised CNN structure for 143-label data.

Layer (type)	Output shape	Param #
Conv2d	(None, 3, 15, 128)	640
Conv2d	(None, 3, 15, 128)	82,048
Conv2d	(None, 3, 15, 256)	164,096
Conv2d	(None, 3, 9, 512)	918,016
Conv2d	(None, 3, 3, 256)	917,760
Flatten	(None, 2304)	0
Dense	(None, 1024)	2,360,320
Dropout	(None, 1024)	0
Dense	(None, 512)	524,800
Dropout	(None, 512)	0
Dense	(None, 512)	262,656
Dropout	(None, 512)	0
Dense	(None, 143)	73,359

Table 1.
The parameters of our 143 classification CNN-DNN model.

after the first layer. Then the previous output bypasses another two convolutional layers with 128 and 256 "1 × 5" filters together with the same paddings. After the first three convolutional layers, the dimension of the data exploded to 3*15*256.

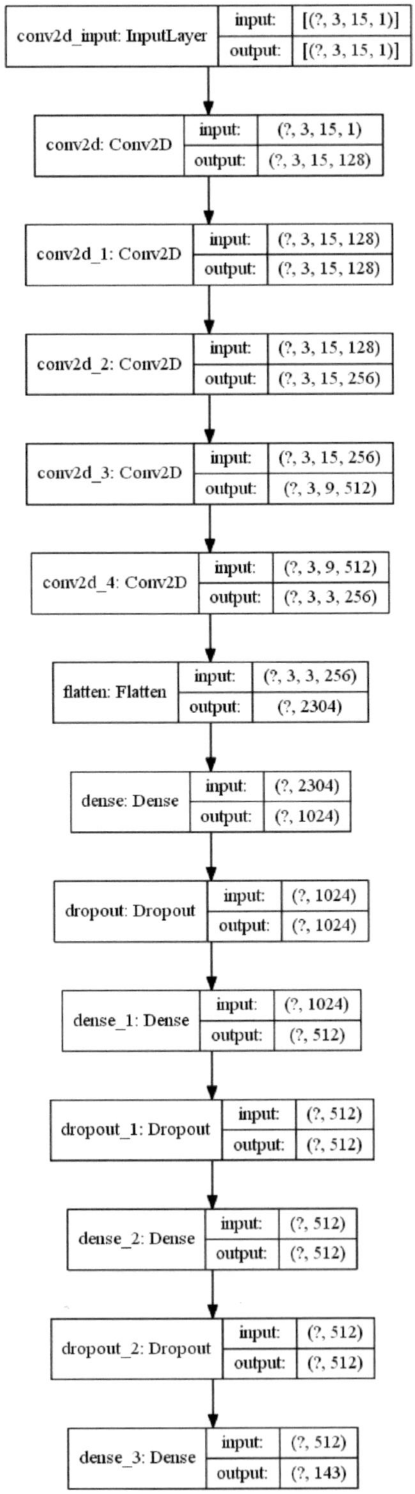

Figure 12.
Model structure for the CNN_DNN network.

Afterwards, another two convolutional layers are added with the same padding methods to downsample the dimensions to 3*3*256 for future operation. The two layers both has 1 × 7 filters with 512, and 256 third dimension respectively. Finally, the 3*3*256 data is flattened to one dimension to bypass three dense layers with 1024, 512, and 512 neurons, respectively. All these layers utilized 30% dropout parameters after dense layers. The employment of the dropout layer is used for solving the overfitting problem during the training process. In order to predict the 143 labels classifications, there is a 143-degree softmax attached at the last layer of the CNN-DNN model. Additionally, the cross-entropy loss function with the Adam Optimizer minimizes the result with the gradient of the next calculation points.

The details of the parameters of the CNN are shown in **Table 1** and **Figure 12**. There are 5,303,695 parameters that need to be trained in this large NN for detecting the different behaviors. In order to train the GTEL CNN much faster and more accurately, a learning rate decay is used during the training process. For the first 75 epochs, the learning rate is $1e^{-4}$, which can learn faster after the weight initializing. After 75 epochs, the exponentially decay of learning rate finds the optimized path for decreasing the loss for the CNN model.

5. Numerical results and discussion

5.1 Training data generation

In order to capture the training data with the two-line elements (TLEs) from the space-track.org, we modified the catalog tracking by the addition of the maneuvers. The different maneuvers indicate different labels in our training data. As an example, the maneuver to increase the energy of the orbital is labeled as 1. On the contrary, the maneuver to decrease the energy of orbital is labeled as 2. In this way, adding no maneuver is labeled as 3. The details related to adding maneuvers are shown as the following:

a. The earth-centered inertial coordinates (ECI) is converted from the TLEs at 0 time step;

b. Use the specified methods (Markov game) shown in Section 3 to propagate the satellites;

c. Convert the 16 waypoints back to azimuth, elevation, range, range rate relative to a ground site.

Therefore, almost 57,332 tracks were generated for training dataset with the other 6371 tracks as the testing dataset. The original first 10,000 tracks are shown in **Figure 13** for 143-space-behavior maneuvers.

The data format is list as:

- Column 1: track id
- Column 2: observation id (from 1 to 15)
- Column 3: Azimuth angle (rad)
- Column 4: Elevation angle (rad)

- Column 5: Range (km)
- Column 6: Training label (from 1 to *143*)

5.2 Results and analysis of our CNN_DNN neural networks

After 1000 epoch optimization, the GTEL systems achieved almost 98% training accuracy with 96% test accuracy as shown in **Figure 14**. There is a little overfitting after 400 epochs of training, which can be solved by adding more dropout layers and other methods. **Figure 15** displays the cross-entropy loss during the training process. The training set loss is always decreasing; however, the loss of validation set is flat after 200 epochs of training, indicating some overfitting issue after 200 epochs of training of the data.

Figure 16 shows the confusion matrix to evaluate our model's performance. As shown, for both training and test datasets, the true labels are almost the same with the predicted label by the GTEL CNN model. The performance evolution pattern is shown in **Figure 17**.

Figure 13.
The first 10,000 training tracks with various space behaviors.

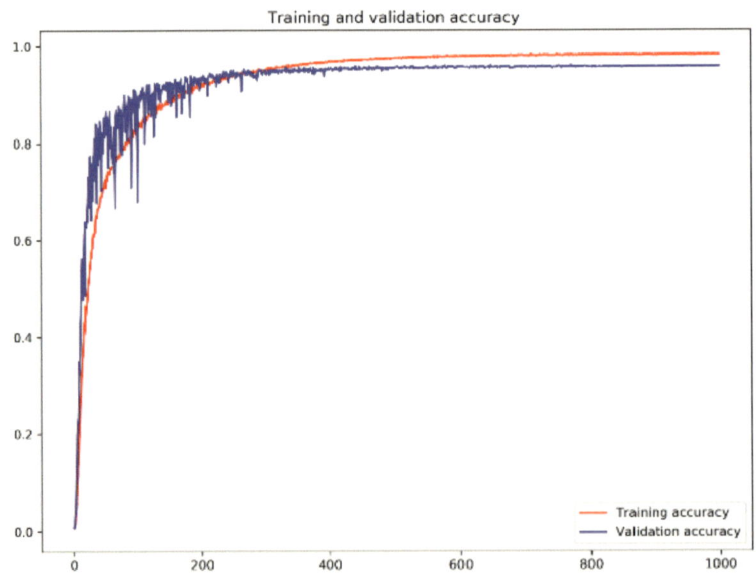

Figure 14.
Training and validation accuracy during the training process.

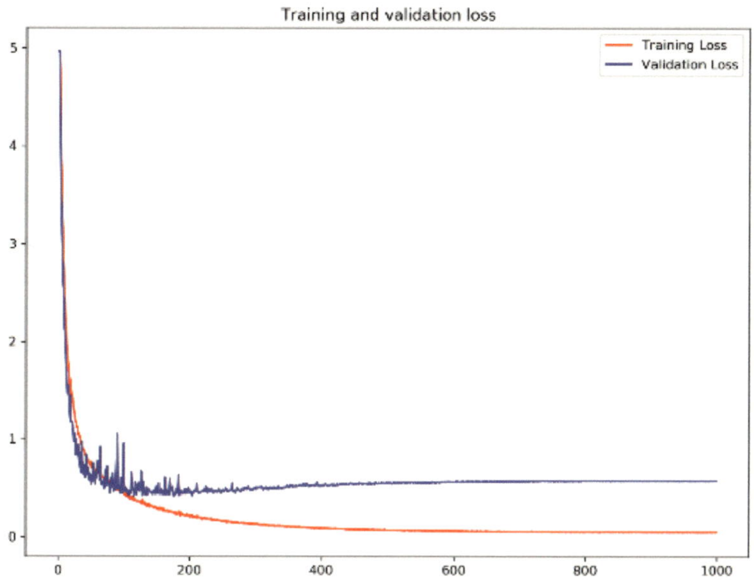

Figure 15.
Training and validation loss during the training process.

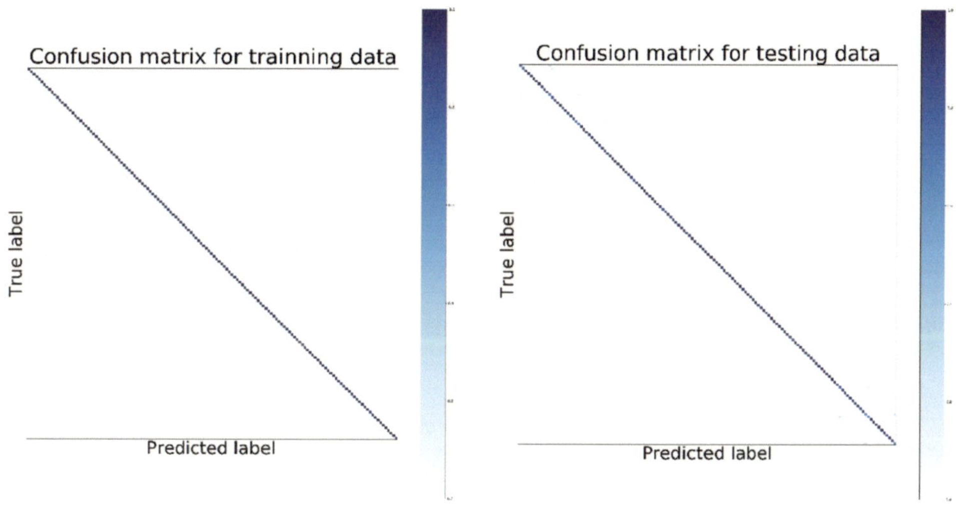

Figure 16.
Confusion matrix of training data and testing data.

For filter-based normalization, changes are grouped normalized by filters, which aims to display the change distributions over iterations for individual filters. For instance, as shown in **Figure 18**, the filter changes in the convolutional layer and dense layers are visualized. The changes are drastic in the first several iterations and become relatively small in the later stages (after 400 epochs) for most of the layers due to learning rate decay as well as the convergence of the GTEL CNN model.

Notice in **Figure 18** (the color maps are shown on the right side of each plot, the whiter the more stable of the training process), there is no constant deep blue color

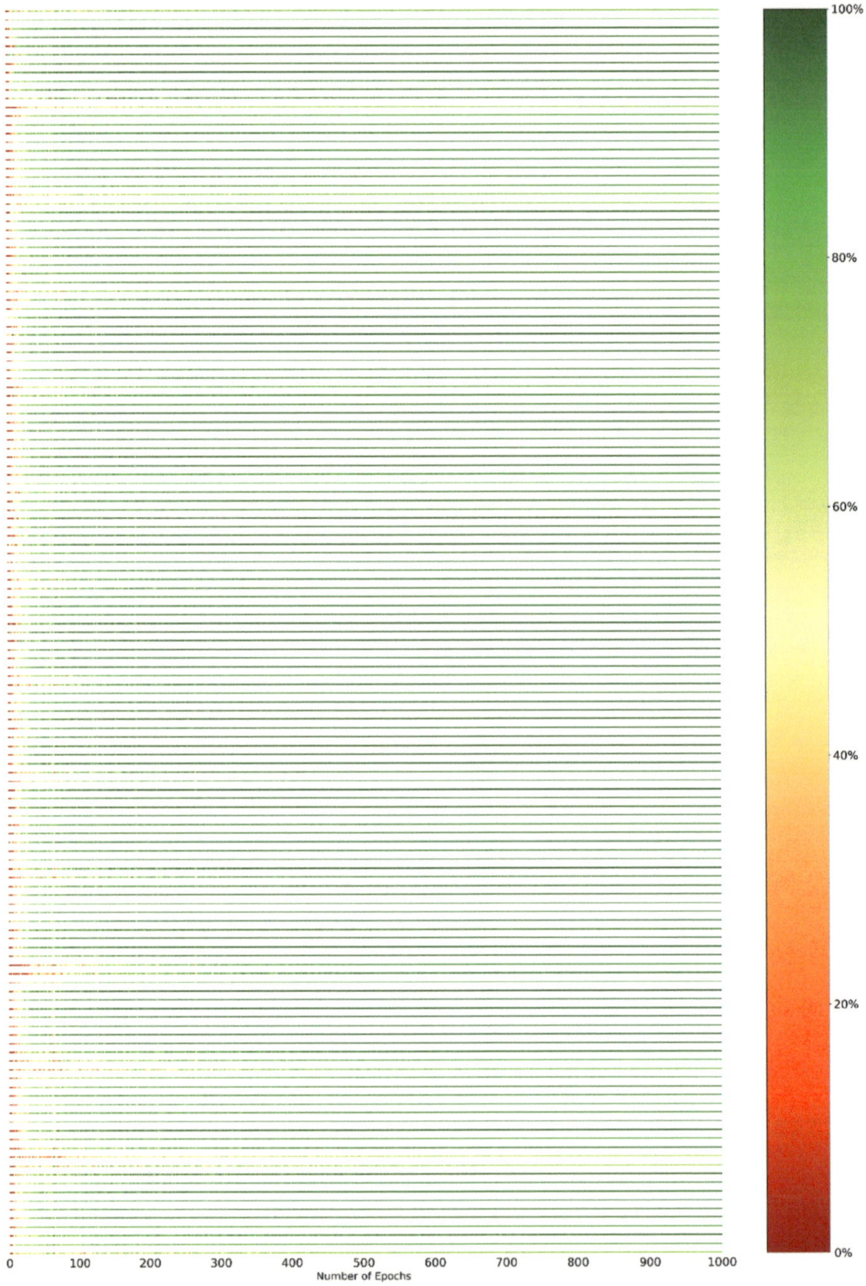

Figure 17.
Overview of validation classes prediction accuracy during training epochs.

for all training process, indicating the great performance and reasonable training configuration during training of the deep learning neural networks. Understanding the meaning for these weights during training process with further checking is very important in the CNN model design, which leads to more reasonable explanation in the deployment of the model.

In addition, we investigated the filter image correlations in a broad overview. As shown in **Figure 19**, rows and columns represent layers and image classes, respectively. A sequential color scheme is used to encode the number of anomaly filters to intuitively represent these relationships between satellite behavior labels and anomaly filters. Using a grid-style visualization, interpretability is possible where rows and columns represent layers and classes of behaviors, respectively. The number of rows and columns equal to the number of layers with anomaly filters and classes with anomaly iterations each class, respectively. In **Figure 19**, the darker the color it is, the more anomaly weights and filters appear in that layer, which are related to that class. From this visualization, it is easy to observe that the second fully connected dense layer has the most anomaly weights and filters among all the other layers. Hence, there are some trends for most anomaly weights and filters during training processes, especially the middle layers of the deep convolutional neural networks. In addition, the anomaly class seems to have the same interval in the dataset, indicating more work is needed for these classes and layers to build a better network. It displays a similar trend compared to the results as shown in **Figure 17**.

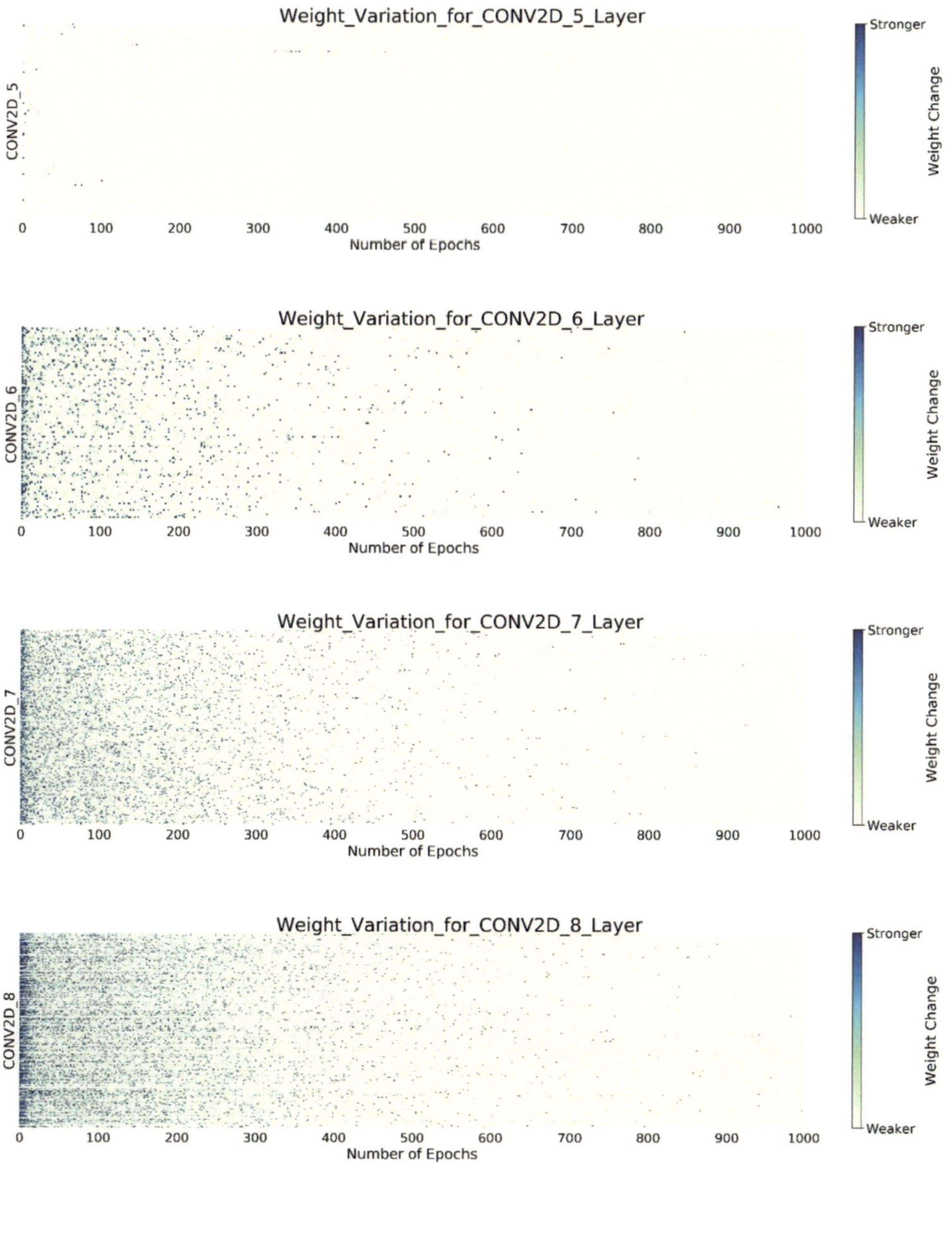

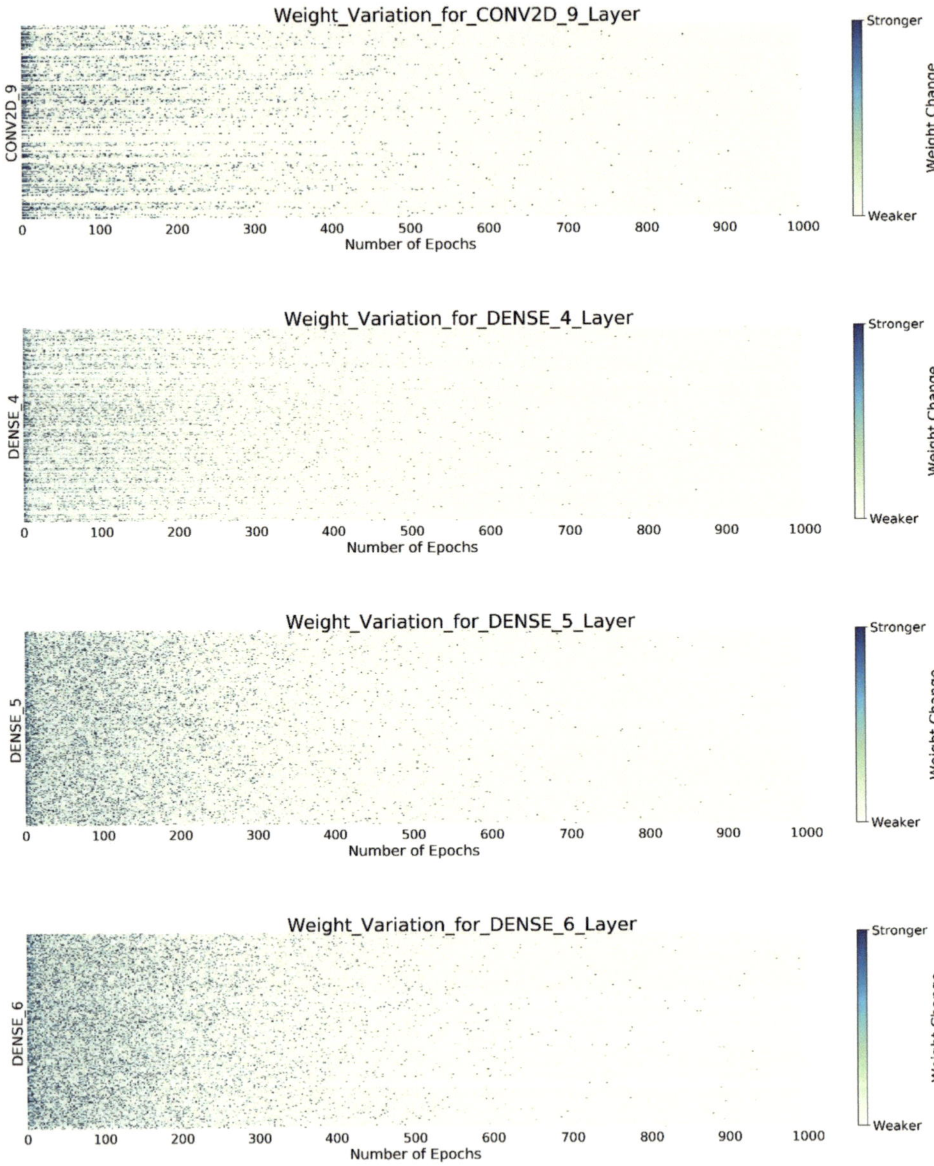

Figure 18.
The weight changes in filters during iterations. More blue color indicates stronger variation for the filters during iterations.

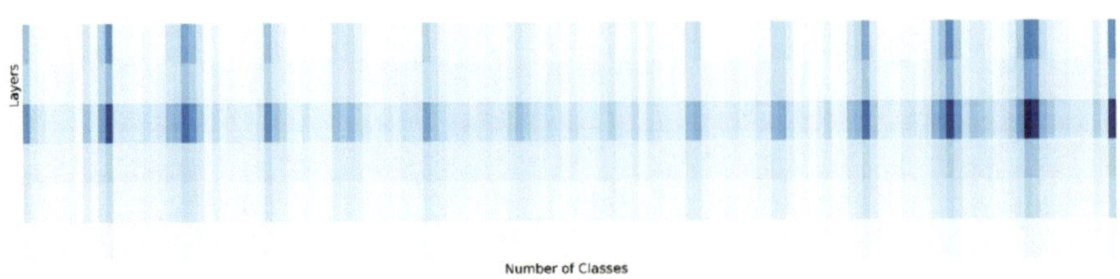

Figure 19.
The abstract version of correlation view.

6. Conclusions

In this chapter, a machine learning design has been presented and implemented to discover space object behaviors. Our GTEL methods using CNN models the circumstance with PE game rather than a control problem. The stochastic modeling/propagation, RSO tracking, and data level fusion are utilized to predict the relations for future space by game reasoning. In order to generate the data for training, the Marko game approach is used with maneuvering strategies. The method provides a way to solve the SSA using unknown behaviors. Additionally, the unknown behaviors exist where a satellite employs the tracking and sensing way to corrupt the tracking estimates to perturb the sensors. On the other hand, the space sensors decrease the uncertainties during tracking process. Finally, the CNN-DNN is used to train the numerical results, where the accuracy is 98% for our classification RSO model with 143 labels.

In the future, a multi-player game theory with adversarial network with SSA will be employed to enhance the deep learning for sensor management, combined tracking, as well as secure communications. Methods for diffusion-based cooperative space object tracking [16] and block chain [17] are emerging as methods for game-theoretical methods.

Acknowledgements

The work was supported under contract FA9453-15-C-0459, FA9453-15-C-0423, and FA8750-19-C-1000. The views and conclusions contained herein are those of the authors and should not be interpreted as necessarily representing the official policies, either expressed or implied, of AFRL, or the U.S. Government.

Author details

Dan Shen[1], Carolyn Sheaff[2], Genshe Chen[1]*, Jingyang Lu[1], Mengqing Guo[1], Erik Blasch[3] and Khanh Pham[4]

1 Intelligent Fusion Technology, Inc., Germantown, MD, USA

2 Information Directorate, Air Force Research Laboratory (AFRL), Rome, NY, USA

3 Air Force Office of Scientific Research (AFOSR), Arlington, VA, USA

4 Space Vehicles Directorate, Air Force Research Laboratory (AFRL), Albuquerque, NM, USA

*Address all correspondence to: gchen@intfusiontech.com

References

[1] Blasch E, Bosse E, Lambert DA. High-Level Information Fusion Management and Systems Design. Norwood, MA: Artech House; 2012

[2] Blasch E. Enhanced air operations using JView for an air-ground fused situation awareness UDOP. In: AIAA/IEEE Digital Avionics Systems Conference, October. 2013

[3] Shen D, Pham K, Blasch E, Chen H, Chen G. Pursuit-evasion orbital game for satellite interception and collision avoidance. Proceedings of the SPIE. 2011;**8044**:89-97

[4] Palatucci M et al. Zero-shot learning with semantic output codes, advances in neural information processing systems. 2009

[5] Mertens JF, Neyman A. Stochastic games. International Journal of Game Theory. 1981;**10**(2):53-66. DOI: 10.1007/BF01769259

[6] Shapley LS. Stochastic games. Proceedings of the National Academy of Sciences of the United States of America. 1953;**39**:1095-1100

[7] Horwood JT, Poore AB. Gauss von mises distribution for improved uncertainty realism in space situational awareness. SIAM/ASA Journal on Uncertainty Quantification. 2014;**2**(1):276-304

[8] Das N, Ghosh RP, Guha N, Bhattacharya R, Mallick B. Optimal transport based tracking of space objects in cylindrical manifolds. The Journal of the Astronautical Sciences. 2019;**66**:582-606

[9] Das N, Deshpande V, Bhattacharya R. Optimal-transport-based tracking of space objects using range data from a single ranging station. Journal of Guidance, Control, and Dynamics. 2019;**42**(6):1237-1249

[10] Shen D, Blasch E, Zulch P, Distasio M, Niu R, Lu J, et al. A joint manifold leaning-based framework for heterogeneous upstream data fusion. Journal of Algorithms and Computational Technology (JACT). 2018;**12**(4):311-332

[11] Jia B, Pham K, Blasch E, Shen D, Wang Z, Chen G. Cooperative space object tracking using space-based optical sensors via consensus-based filters. IEEE Transactions on Aerospace and Electronic System;**52**(4):1908, 2016-1936

[12] Chen G, Shen D, Kwan C, Cruz J, Kruger M, Blasch E. Game theoretic approach to threat prediction and situation awareness. Journal of Advances in Information Fusion. 2007;**2**(1):35-48

[13] Shen D, Jia B, Chen G, Pham K, Blasch E. Game optimal sensor management strategies for tracking elusive space objects. In: Proceedings of IEEE Aerospace Conference. 2017:1-8. DOI: 10.1109/AERO.2017.7943676

[14] Zou H, Lang H, et al. Covert photo classification by deep convolutional neural networks. Machine Vision and Applications. 2017;**28**(5-6):623-634

[15] Available from: https://www.tensorflow.org/

[16] Jia B, Pham K, Blasch E, Shen D, Chen G. Diffusion-based cooperative space object tracking. Optical Engineering. 2019;**58**(4):041607

[17] Xu R, Chen Y, Blasch E, Chen G. An exploration of Blockchain-enabled decentralized capability-based access control strategy for space situation awareness. Optical Engineering. 2019;**58**(4):041609

Section 3

Satellite Systems Design

Chapter 5

Future Satellite System Architectures and Practical Design Issues: An Overview

Tien M. Nguyen

Abstract

This chapter discusses existing and future trends on the design and build of "Modular" and "Open" satellite Bus and mission payload along with practical design issues associated with the use of Modular Open System Approach (MOSA). Existing modular Bus and mission payload architectures for typical commercial, civilian, and military satellite systems will be discussed. The chapter provides space industry views on "Open" versus "Close" interfaces design and addresses the challenges associated with open interfaces using Open System Architecture (OSA) approach using MOSA principles. The system interfaces discuss in this chapter include (i) internal to satellite Bus and mission Payload (PL), (2) between satellite Bus and mission payload, and (3) external to both satellite Bus and mission payload.

Keywords: open system architecture, Modular Open System Approach (MOSA), satellite Bus, mission payload (PL), modular architecture, open Interface, close Interface, modular satellite Bus, modular Mission payload

1. Background and introduction

Typical commercial and civilian satellite systems take about 2–3years to build and launch [1–4], while military systems take between 7 and 10 years [5, 6]. A typical production flow for assembling and launching of a space vehicle is presented in Ref. [6] and redrawn in **Figure 1** as introduction steps for better understanding of the design, build and launch of a satellite system. This chapter focuses on practical design issues for satellite Bus' and mission PL's system/subsystem components builds, and corresponding interface-design's challenges associated with satellite Bus integration, mission PL integration, and satellite system integration. A survey of existing commercial, civilian and military satellite systems revealed that a typical satellite Bus includes the following modular components [7–16]:

- Bus Subsystem 1—Bus TX/RX Antenna Subsystem (BAS): Provide Bus's Receive (RX)/Transmit (TX) antennas and associated Bus's antenna beam control functions;

- Bus Subsystem 2—Bus Communication RF Front-End-Back-End Subsystem (BCom-RFS): Provide Low Noise Amplifier (LNA), High Power Amplifier (HPA), satellite Bus Down/Up Radio Frequency (RF)-to-Intermediate

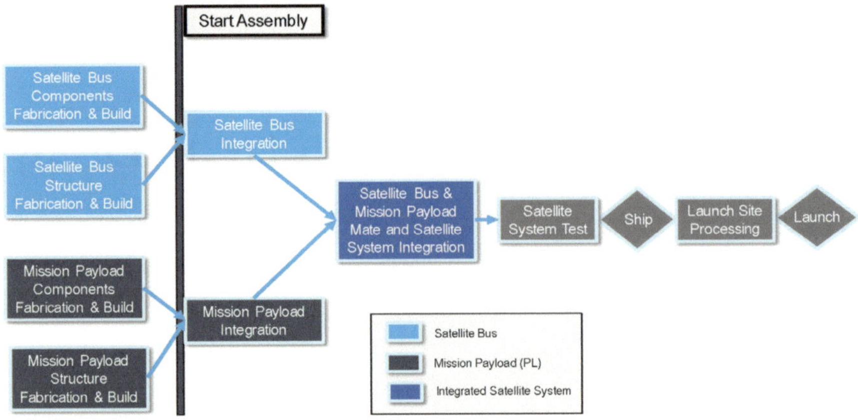

Figure 1.
A typical satellite system production flow (redrawn from [6]).

Frequency (IF) conversion, and Analog-to-Digital/Digital-to-Analog conversion functions—Note that typical HPAs are Traveling Wave Tube Amplifier (TWTA) and Solid State Power Amplifier (SSPA), and some advanced satellite transponders use Linearized TWTA (L-TWTA) or L-SSPA in the RF Back-End Subsystem;

- Bus Subsystem 3—Bus Command & Data Handling Subsystem (BC&DHS): On-board computer that interfaces with all Bus components;

- Bus Subsystem 4—Bus Telemetry-Tracking & Command Subsystem (BTT&CS): Process uplink satellite Bus command data, perform satellite tracking functions and provide downlink Bus telemetry reporting satellite Bus's heath and conditions;

- Bus Subsystem 5—Bus Electrical Power Subsystem (BEPS): Provide and regulate Bus power;

- Bus Subsystem 6—Bus Thermal Control Subsystem (BTCS): Maintain Bus' thermal environments;

- Bus Subsystem 7—Bus Altitude and Determination Control Subsystem (BADCS): Provide satellite stabilization, control and positioning;

- Bus Subsystem 8—Bus Propulsion Subsystem (BPS): Provide propulsion functions for satellite maneuvering;

- Bus Subsystem 9—Bus Communication Security Subsystem (BCOMSEC): Provide Bus data encryption and decryption functions to protect data from intruders. Typically, BCOMSEC is tightly coupled with BTT&CS;

- Bus Subsystem 10—Bus Structure & Mechanism Subsystem (BS&MS): Provide structure and mechanism to mount all satellite Bus components.

Similarly, our survey also revealed that a typical mission PL consists of the following modular components [7–16]:

- PL Subsystem 1—PL PAS: Similar to BPAS but for mission PL;
- PL Subsystem 2—PL Com-RFS: Similar to BCom-RFS but for mission PL;
- PL Subsystem 3—PL Digital Processing Subsystem (PDPS): Provide mission specific processing functions. For SATCOM missions, specific processing functions can be dynamic resources control, channelization processing, etc. For PNT missions, the functions can be time transfer processing functions. For imaging/sensing missions, the functions can be image preprocessing functions;
- PL Subsystem 4—PL C&DHS: Similar to BC&DHS but for mission PL;
- PL Subsystem 5—PL TT&CS: Similar to BTT&CS but for mission PL;
- PL Subsystem 6—PL EPS: Existing PLs use power supply from satellite BEPS;
- PL Subsystem 7—PL TCS: Maintain PL's thermal environments;
- PL Subsystem 8—PL ADCS: Existing PLs use ADCS from the satellite BADCS;
- PL Subsystem 9—PL PS:: Existing mission PLs use PS from the satellite BPS;
- PL Subsystem 10—PL COMSEC: Similar to BCOMSEC but for mission PL;
- PL Subsystem 11—PL Frequency & Timing Subsystem (PFTS): Provide reference frequency and timing functions to meet specific mission requirements;
- PL Subsystem 12—PL Transmission Security Subsystem (TRANSEC): Provide security functions to combat unintentional and/or intentional Radio Frequency Interference (RFI) (e.g., frequency hopping/de-hopping, frequency spreading/de-spreading);
- PL Subsystem 13—PL Specific Mission Suite (SMS): Provide specific mission PL processing functions depending on whether the mission is Satellite Communications (SATCOM) mission or Position Navigation and Timing (PNT) mission or Imaging/Sensing mission [7–16];
- PL Subsystem 14—PL Structure & Mechanism Subsystem (BS&MS): Provide structure and mechanism to mount all mission PL components.

In practice, the above satellite Bus modular components can be found in the following typical satellite Busses [7, 11, 14]:

- Loral Satellite Bus 1300 or Loral 1300
- Lockheed Martin (LM) A2100A/AX-Land Mobile/AX-High Power
- Boeing 702HP/HP-GEM/MP/702SP and 502.

For achieving optimum weight and power, existing satellite Bus and mission PL are tightly coupled together with customized interface design. The industry trends for the design and build of future satellite systems are moving toward OSA using MOSA principles, in which the satellite Bus is loosely coupled with the mission PL using "Open" and widely accepted interface standards. The key communication linkage between a satellite Bus and a mission PL is the communication data Bus. Currently, majority of satellite Busses employ the standard 1553 data Bus for data communications among Bus components, and between the satellite Bus and mission PL components. The communications over 1553 data Bus is limited to 1 Mega bit per second (Mbps). Recently, there was an advanced development effort that was funded by the U.S DOD to develop new 1553 standards called 1553 Enhanced Bit Rate (EBR–1553) increasing the speed to 10 MB/s [17]. The EBR-1553 requires a star/hub topology to provide the higher data rate and additional components to implement the architecture. For data rates larger than 10 Mbps, space industry trend is moving toward SpaceWire data Bus that was recently developed in Europe for use in commercial satellites and scientific spacecraft [18].

The objective of this chapter is three-fold: (1) Provides an overview of existing modular satellite Bus, mission PL architectures and related communication data Busses, (2) Discusses future trends on the modular and open design and build of satellite Bus and mission payload using MOSA principles, and (3) Addresses the practical design challenges associated with "Modular" and "Open" design for future satellite Bus and mission PL. The chapter is organized as follow: (i) Section 2 describes existing modular satellite Bus and mission PL architectures and related communication data Busses; (ii) Section 3 presents industry view on "Open" and "Close" interfaces for connecting satellite system components and existing popular standards; (iii) Section 4 discusses the interface design challenges and provides an overview of MOSA and related DOD Guidance and assessment tools for MOSA implementation; (iv) Section 5 provides examples how to transition modular satellite Bus and mission PL architectures to modular-and-open architectures using MOSA implementation approach and tools in Section 4; and (v) Section 6 concludes the chapter with remarks on the benefits associated with the proposed approach.

2. Existing satellite systems, related interfaces and standards

Figure 2 describes an overview of existing satellite systems, consisting of a satellite Bus, a mission PL and a typical set of interfaces between the Bus and PL using a standard data Bus. A typical set of interfaces between the satellite Bus and a mission PL includes seven interface types, namely: (i) Physical & Mechanical Interface, (ii) Electrical/Power/Cable Interface, (iii) Grounding Interface, (iv) Software & Data Interface, (v) Electromagnetic Compatibility (EMC)/ Electromagnetic Interference (EMI)/Electromagnetic Pulse (EMP) and Electro Static Discharge (ESD) Interface, (vi) Thermal Interface, and (vii) Frequency & Timing (F&T) Interface. This section focuses on satellite Bus and mission PL architectures and the data interfaces between them. Subsections 2.1 and 2.2 describe existing satellite Bus and mission PL architectures along with related interfaces and industry standards, respectively. Subsection 2.3 discusses existing standard 1553 data Bus and the pushes from space industry moving toward military standard 1553-B data Bus (MIL-STD-1553-B) and high-data-rate SpaceWire data Bus.

2.1 Existing satellite bus, related interfaces and standards

As described in Section 1, existing satellite Bus architecture includes typical 10 modular components, namely, BAS, BComRFS, BC&DHS, BTT&CS, BEPS, BTCS, BADCS, BPS, BCOMSEC and BS&MS. A functional description for each of these modular Bus components is also described in Section 1. **Figure 3** illustrates a notional block diagram for existing modular satellite Bus architecture. The figure

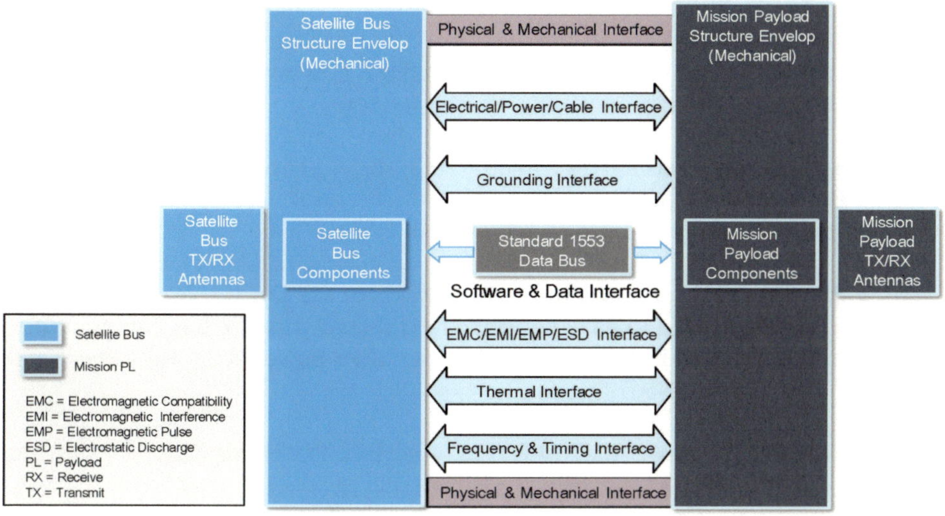

Figure 2.
Overview of existing satellite Systems using standard 1553 data bus.

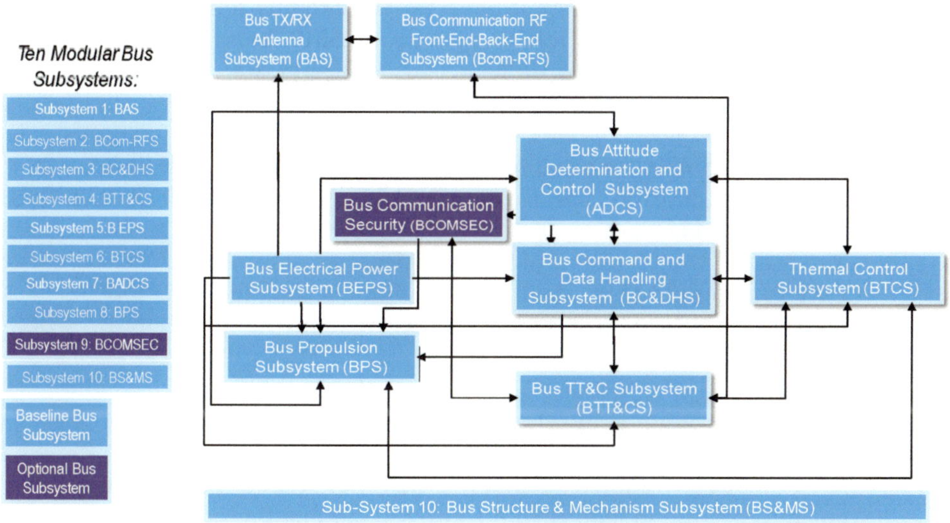

Figure 3.
Existing notional modular satellite bus architecture.

shows that space industry has used the modular design concept to architect the satellite Bus, where common functions are group together and then isolate or separate from the other group of functions. As an example, BAS consists of a group of antenna components and control functions (e.g., antenna pointing, beamforming, etc.), which is separated and isolated from BComRFS. It is important to note that the figure also shows how these satellite Bus components are connected together, i.e., the lines with arrows connecting them. These lines represent the interfaces among the Bus components, where the interface can be any of the seven interface types described above. Below is a list of some of the existing interfaces and associated standards for existing satellite Bus based on National Aeronautical and Space Administration (NASA), European Space Agency (ESA), U.S. DOD and international Consultative Committee for Space Data System (CCSDS) standards [19–26]:

- Typical NASA Electrical/Power/Cable Interface Standards [19, 20]:
  - Satellite Bus shall protect its own electrical power system via overcurrent protection devices on its side of the interface.
  - Satellite Bus shall deliver a maximum transient current on any Power Feed Bus of 100% (that is, two times the steady state current) of the maximum steady-state current for no longer than 50 ms.
  - Bus Survival Heaters, which are elements of the Bus thermal subsystem, shall be required to have power to heat certain satellite Bus components during off-nominal scenarios when the BEPS power is not fully energized.
- Typical U.S. DOD EMC/EMI/EMP/ESD Interface Standards [21]:
  - Power line conducted emissions for satellite Bus equipment shall meet the EMC interface specification specified in SMC Standard Handbook, SMC-S-008, Section 6, 6.01, 6.02, 6,03, 6.04, 6.05, 6.06, 6.07, and 6.08.
  - Power line conducted susceptibility for satellite Bus equipment shall meet the EMC interface specification specified in SMC Standard Handbook, SMC-S-008, Sections 6, 6.10, 6.11, 6.12, 6.13, 6.14, 6.15, 6.16, 6.17, 6.18 and 6.19.
  - ESD susceptibility for satellite Bus equipment shall meet the EMC interface specification specified in SMC Standard Handbook, SMC-S-008, Section 6, 6.43.
  - EMP susceptibility for satellite Bus equipment shall meet the EMC interface specification specified in SMC Standard Handbook, SMC-S-008, Section 6, 6.45.
- Typical NASA Grounding Interface Standards [20, 22]:
  - Satellite Bus EPS should ground in a way that reduces introducing stray currents or ground loop currents into the satellite Bus components.
  - Satellite Bus ground interface shall follow NASA single-point ground or multiple-point ground architecture.
- Typical NASA Thermal Interface Standards [19, 20]:
  - A conductive heat transfer of 15 W/m2 or 4 W shall be considered small enough to meet the intent of being thermally isolated.
- Typical Software & Data Interface Standards [19, 20, 23, 24, 25, 26]:
  - Satellite Bus command and telemetry data formats shall be NASA Unified S-Band (USB)/CCSDS standards or U.S. DOD Space-Ground Link Subsystem (SGLS) standards. Note that (i) most of NASA and ESA standards are CCSDS compliance for interoperability purpose, and (ii) some military systems have both USB and SGLS capabilities.

- Satellite Bus "Safe Mode" is a combined satellite Bus components hardware and software configuration that shall be designed to protect the components from possible internal or external harm while making minimal use of satellite Bus resources (e.g., power).

- Satellite Command SAFE Mode shall be required to protect and preserve satellite Bus components under anomalous and resource constrained conditions.

- Satellite Bus components shall respond to uplink commands from the Satellite Operation Center (SOC) to suspend and resume the transmission of the Components' telemetry data. For commercial satellite systems, SOC can also control the mission PL.

For military applications, majority of satellite Busses are usually designed using contractor's custom designed interfaces and very tightly couple together to reduce weight, size and power. It is for this reason, current military satellite BTT&CS component also include the COMSEC component. For commercial applications, satellite developers are also concerned with weight, size and power reduction, but they are also concerned with component refresh and upgrade without redesigning the satellite Bus, hence commercial satellites tend to use modular Bus components and widely accepted interface standards to connect the internal Bus components. Industry views on the "open" and "close" interfaces will be addressed in Section 4.

2.2 Existing mission payload, related interfaces and standards

As pointed out in Section 1, existing mission PL architecture consists of 14 modular components, but there are three PL components that rely on the satellite Bus' design, namely, PL EPS, PL ADCS and PL PS. Therefore, the mission PL architecture usually has 11 modular components, including PL AS, PL Com-RFS, PDPS, PL C&DHS, PL TT&CS, PL TCS, PL COMSEC, PFTS, PL TRANSEC, PL SMS and PL S&MS. A functional description for each of these mission PL modular components is also provided in Section 1. **Figure 4** presents a notional block diagram for existing modular mission PL architecture. Similar to the satellite Bus

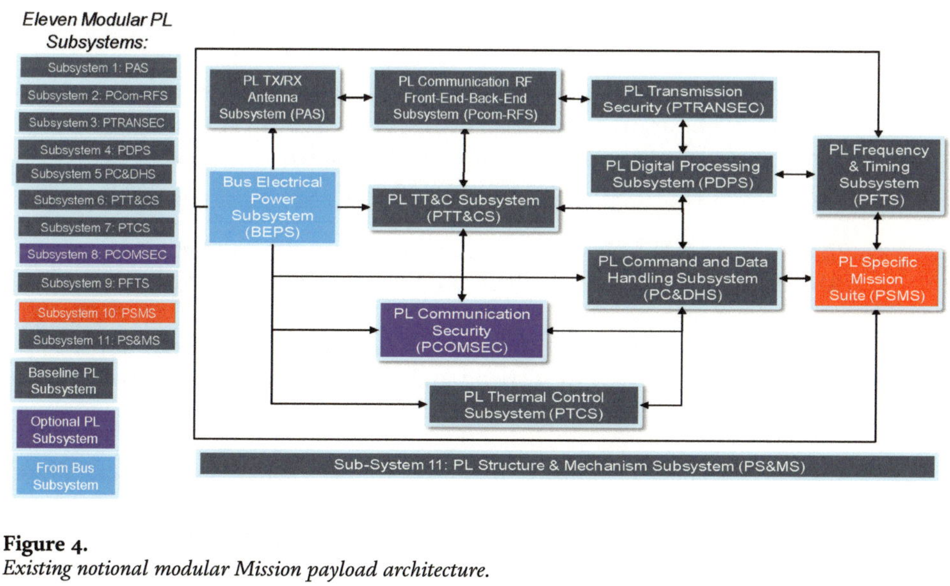

Figure 4.
Existing notional modular Mission payload architecture.

design, the space industry has also applied the modular design concept to architect the mission PL. Below is a list of some of the existing interfaces and associated standards for existing mission PL leveraged from NASA, ESA, U.S. DOD and international CCSDS standards [19–26]:

- Typical NASA Electrical/Power/Cable Interface Standards [19, 20]:
    - Sizing all components of the mission PL power harness, such as the wires, connectors, sockets, and pins to the peak power level shall be required by the mission PL equipment in addition to satellite Bus to prevent damage to the power harnessing.
    - PL Survival Heaters shall be required to have power to heat certain mission PL components during off-nominal scenarios when the BEPS power is not fully energized.
- Typical U.S. DOD EMC/EMI/EMP/ESD Interface Standards [21]: Similar to satellite Bus discussed above but for mission PL.
- Typical NASA Grounding Interface Standards [20, 22]: Similar to satellite Bus discussed above but for mission PL.
- Typical NASA Thermal Interface Standards [19, 20]:
    - The mission PL thermal design should be decoupled from the satellite Bus at the mechanical interface between the satellite Bus and neighboring mission payload to the maximum practical extent.
    - A conductive heat transfer of 15 W/m2 or 4 W shall be considered small enough to meet the intent of being thermally isolated.
- Typical Software & Data Interface Standards [19, 20, 23, 24, 25, 26]:
    - Mission PL command and telemetry data formats shall be NASA USB/CCSDS standards commercial applications or U.S. DOD SGLS standards for military applications. Some military systems have both USB and SGLS capabilities.
    - PL "Safe Mode" is a combined mission PL components hardware and software configuration that shall be designed to protect the PL components from possible internal or external harm while making minimal use of satellite Bus resources (e.g., power).
    - PL Command SAFE Mode shall be required to protect and preserve mission PL components under anomalous and resource constrained conditions.
    - Mission PL components shall respond to uplink commands from Mission Control Center (MCC) to suspend and resume the transmission of the mission PL components.
    - Mission PL shall be responsible for on-board mission data storage capabilities.

For most commercial applications, the MCC can be merged with the SOC, and the mission PL TT&CS (PTT&CS) and PL CD&HS (PCD&HS) components can be incorporated into satellite (i) Bus TT&C (BTT&CS) and (ii) Bus CD&HS (BCD&HS) components, respectively. Similar to the satellite Bus interfaces design, for military applications, the mission PL components are tightly coupled using contractor's custom designed interfaces. For commercial applications, the mission PL components are loosely coupled using widely accepted open interfaces.

2.3 Existing data busses

Subsections 2.3.1 and 2.3.2 provide an overview of standard 1553 and SpaceWire communication data Busses, respectively.

2.3.1 Standard 1553 data bus

Existing commercial, civilian and military satellite data Busses have been using Military Standard 1553B (MIL-STD-1553B) data Bus for communications among satellite Bus and mission PL components. **Figure 5** describes a typical MIL-STD-1553B System [17, 27, 28]. This figure uses MIL-STD-1553B terminologies: (i) the Bus Controller (BC) is considered as an Intelligent Terminal (IT) that is located in the satellite mission computer, which is usually referred to as a Satellite Bus C&DH component, and (ii) Remote Terminal (RT) is considered as a slave terminal that is located in satellite platform components, which can be located in any satellite Bus or mission PL components.

Figure 5 shows a typical commercial satellite system with RTs located in both satellite Bus and mission PL components. As an Example, the RTs located in satellite components are BAS, BADCS, BTCS and BTT&CS; and RTs located in the mission PL components are PAS, PTCS, PDPS, PTRANSEC, and PFTS. For military applications, the Mission Computer (MC) can be located in both satellite Bus and mission PL, where the MC in the satellite Bus is responsible for all control functions associated with the satellite operations and MC in the mission PL is responsible for all control functions related to the mission PL operations.

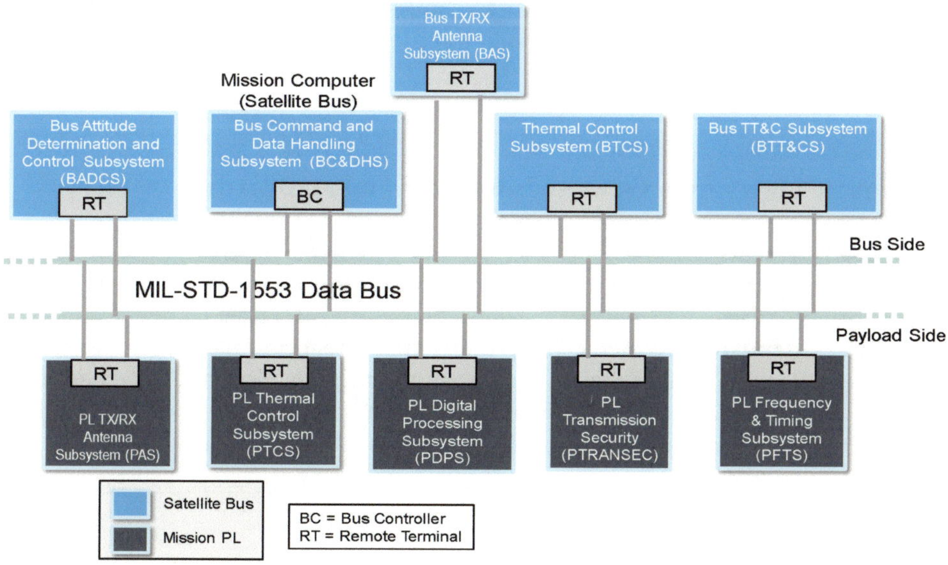

Figure 5.
Typical civilian and commercial MIL-STD-1553B satellite Systems.

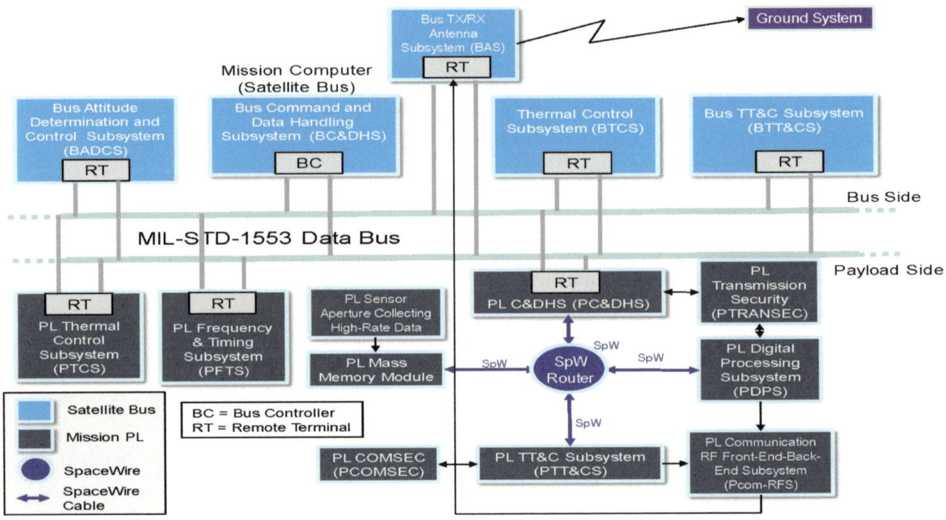

Figure 6.
Typical civilian and commercial SpaceWire satellite Systems.

2.3.2 Standard SpaceWire data bus

SpaceWire (SpW) is an industry standard with protocol derived from IEEE-1355 and ECSS-E50-12C managing by the international SpW working Group [18, 29, 30]. The SpW standard is a self-managing serial protocol that provides a high-speed data rates from 2 to 400 Mbps, and low power serial interface using LVDS[1] Drivers with distances up to 30 feet while offering a flexible simple user interface. **Figure 6** illustrates typical uses of SpW data Bus with a PCD&HS, a SpW Router and SpW cables for connecting mission PL components. Some examples of existing satellite programs employed SpaceWire standard are: TacSat (part of the U.S. Operationally Responsive Space Program), NASA Lunar Reconnaissance Orbiter (Orbiting the Moon taking high resolution images), ESA Sentinel-3 (a pair of satellites providing operational Earth observation services using optical and microwave instruments), and Japanese NEC NEXTTAR (one of the first spacecraft designed using SpW for all of its onboard communications).

3. Industry view: open vs. close interfaces and standards

Figure 7 presents the space industry view on open and close interface design. This view separates the interface design into two categories, namely, Contractor Proprietary Interface and Contractor Non-Proprietary Interface. Under this view, the interface standards are then classified into two categories, namely, Preferred and Non-Preferred Interface Standards. Based on this view, Section 3.1 defines open interface design, and Section 3.2 defines close interface design. Section 3.3 provides a list of existing popular open standards widely accepted by space industry.

---

[1] LVDS is defined as Low Voltage Differential Signaling TIA/EIA-644, is a technical standard that specifies electrical characteristics of a differential, serial communication protocol. LVDS Drivers use 80% less current than current popular Pseudo Emitter-Coupled Logic (PECL) devices.

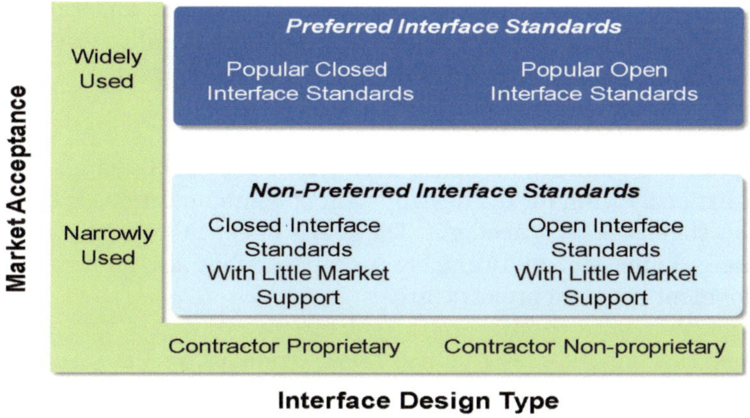

Figure 7.
Industry view on open and closed interfaces design.

3.1 Open Interface design

From **Figure 7**, the open interface design falls into the contractor non-proprietary design category. For the interface design to be open, the interface design shall not be contractor proprietary and that the interface shall use either popular open interface standards widely accepted by space industry or open interface standards with little market support and narrowly used by space industry. Thus, a popular open interface design is a non-proprietary design that uses popular open interface standard that is widely used by space industry. The benefits of open interface design for the satellite buyers are (i) improving competition allowing various space vendors (or contractor) to build open satellite Bus and mission PL subsystem components, (ii) ease of refresh and technology upgrade allowing to swap subsystem components without impacting the overall system, (iii) ease of adapting to new requirements and operational threats, (iv) incorporating innovation by allowing operational flexibility to configure and reconfigure a mission PL quickly to meet rapidly changing operational requirements, (v) enabling cost saving and cost avoidance during the design and sustainment phases by reusing technology and Software/Hardware/Middleware (SW/HW/MW) components, and using existing standardized HW/SW/MW parts and modules, and (vi) improving interoperability where severable HW/SW/MW modules can be changed independently.

3.2 Close interface design

As shown in **Figure 7**, the close interface design shall fall into contractor proprietary category. For an interface design to be close, it shall be contractor proprietary and that the interface shall use either close interface standards with little market support narrowly used by space industry or popular closed interface standards widely used by space industry. Thus, a popular close interface design is a contractor proprietary design that uses popular closed interface that is widely used by space industry. The key benefits of close interface design are the potential reduction of weight, size, power and manufacturing cost.

3.3 Popular open standards

Based on **Figure 7**, the criteria for popular open standards are (i) publicly available and widely used by both satellite Bus and mission PL vendors, (ii) community and/or industry consensus-based that are matured and stable, and (iii) technically adequate for all future commercial, civilian and military satellite systems. Following is a list of current popular standard organizations and widely adopted open standards [18–31]:

- Consultative Committee for Space Data Systems (CCSDS) Standards: is a multi-national forum for the development of communications and data systems standards for spaceflight. The goal is to enhance governmental and commercial interoperability and cross-support, while also reducing risk, development time and project costs.

- AIAA Space Plug-and-play Avionics (SPA) Standard: SPA is a set of AIAA standards developed for spacecraft platform, subsystem, and component (including payload) developers for integrating plug-and-play characteristics into spacecraft structures, avionics, and hardware and software components to promote their rapid integration. The SPA community anticipates adding protocols (e.g., Ethernet as SPA-E) as the PnP capabilities are normalized.

- MIL-STD-1553 Standard: is a military standard published by the United States Department of Defense that defines the mechanical, electrical, and functional characteristics of a serial data Bus.

- SpaceWire Standard: is a spacecraft communication network standard based in part on the IEEE 1355 standard of communications. It is coordinated by the European Space Agency (ESA) in collaboration with international space agencies including NASA, Japanese Space Agency (JAXA) and Russian Federal Space Agency (RKA).

- NASA/SMC/Aerospace Hosted Payload Interface Design (HPID): this design guideline provides a prospective Instrument Developer with technical recommendations to assist them in designing an Instrument or Payload that may be flown as a hosted payload on commercial satellites flown in Low Earth Orbit (LEO), or Geostationary Earth Orbit (GEO).

- SQL for databases specified in ANSI ISO/IEC 9075–1, ISO/IEC 9075–2, ISO/IEC 9075–3, ISO/IEC 9075–4, ISO/IEC 9075–5.

- HTML for presentation layer specified in XML 1.0 www.webstandards.org.

- XML for data transfer.

- Web Services for remote system calls.

- U.S. Space and Missile Systems Center (SMC) approved a project for developing Common Payload Interface Specification (CoPaIS) standard for satellite-to-payload Command and Data Handling (C&DH) interface intended for all future SMC procured medium to large satellites [31].

- Other popular standards: MIL-STD-1553B, CAN Bus, RS-422 (TTC-B01 Protocol)/EIA/TIA-422, RS-422 (PC-Protocol).

4. Interfaces design challenges and MOSA implementation

This section addresses the design challenges and is divided into four subsections, including: (i) Subsection 4.1 discusses interface design and practical design issues; (ii) Subsection 4.2 introduces MOSA concept; (iii) Subsection 4.3 presents DOD MOSA guidance and the U.S. Naval Open Architecture (NOA); and (iv) Subsection 4.4 discusses MOSA tools and approach for MOSA implementation that addresses the design issues identified in Subsection 4.1.

4.1 Interface design and practical design challenges

The interfaces between satellite subsystem components can be SW, HW or MW interfaces. The design and build of these interfaces are well incorporated into any satellite subsystem components "Design Product" and associated "Design Process".

The Design Product includes System Architecture, Interface Product, Independent Verification and Validation (IV&V) Test Plan, Schedule, Design Approach, Acceptance Criteria, and System-Built Product.

For MOSA, the Design Process is expected to incorporate MOSA into: Architecture Process, Interface Management, IV&V Process, and System Engineering and Integration (SE&I) Process. The Interface Product and its open interface design using MOSA along with the Interface Management are the key challenges in the development of open-and-modular satellite systems. The key design challenges for the design and build of open-and-modular satellite systems are:

- Challenge 1: Determination of Key Open Subsystem (KOSS): This is also known as KOSS Selection. Ideally, all modular subsystem components should be made open. But this is not practical, because some interfaces need to be customized using close interface design due to weight, size and power reduction requirements. The key challenge here is to identify a set of criteria that can be used for KOSS selection. Subsections 4.3 and 4.4 will address this challenge.

- Challenge 2: Designation of Key Interfaces for the Selected KOSS: Satellite system designers need to identify a subset of selected set of KOSS components that can be designated as key interfaces. The key challenge here is to identify a business case for the designated key interfaces. Subsections 4.3 and 4.4 will describe selection criteria and tool to address this challenge.

- Challenge 3: Selection of Open Standards for the Designated Key Interfaces: Selection of the popular and open standards for the designated key interfaces is also a potential challenge for the designers. The selection should be based on the cost, required technical specification and availability of "open" products in the market and their usage by space industry. SubSection 3.3 above provides a list of some existing open and popular standards. Subsection 4.4 will discuss how to resolve this challenge by developing a business case to justify the selection of key interfaces and associated open standards.

- Challenge 4: Management of Key Open Interfaces: Not all identified key interfaces in Challenge 2 can be designated an open interface standard at the initial system design phase due to market unavailability. Hence, managing these interfaces can be a potential challenge ensuring that they will be "open"

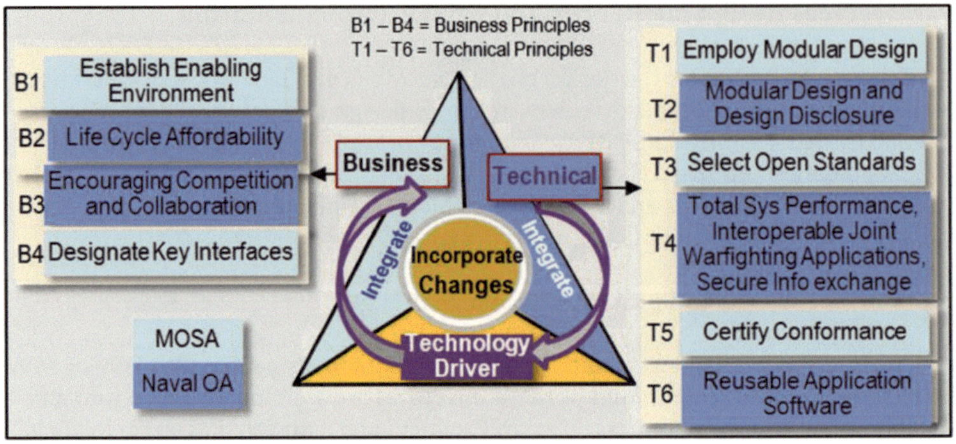

Figure 8.
MOSA and U.S. NOA approach.

by the time of Full Operational Capability (FOC) deployment. Section 4.3 discusses DOD guidance for managing key interfaces for military satellite system development.

4.2 Introduction to MOSA

U.S. DOD recommends OSA design using MOSA principles for future military satellite system development with a goal to achieve a balance between business and technical objectives that make a business sense in terms of (i) increase competition and lower system acquisition cost, and (ii) lower sustainment cost over its life cycle. MOSA design approach requires to implement five MOSA principles, including two Business (B) and three Technical (T) principles [1, 2]. **Figure 8** captures these five B and T principles. Recently, U.S. Navy augmented MOSA principles with addition five Naval Open Architecture (NOA) principles, including two business and three technical principles as shown in **Figure 8** [3].

Subsections 2.1 and 2.2 provide current implementation of the Technical principle 1 (T1) for the modular design of satellite Bus and mission PL, respectively. The remaining Subsections 4.3 and 4.4 discuss the implementation of T2, T6, Business principle 3 (B3), and B4 using DOD guidance for addressing the challenges presented in Subsection 4.1.

4.3 DOD guidance

MOSA mandated the space system technical requirements be based on the maximum extent practicable on open standards as indicated in Section 3.2 of the U. S. DOD Guidebook for Program Managers [1]. The book provides MOSA[2] guidelines and contract language for generating a Request for Proposal (RFP) [1]. At the minimum, the RFP shall incorporate the following MOSA tasks that can help to minimize the MOSA implementation risk in the design, build and test of new satellite systems:

---

[2] Note that the term Open System Architecture (OSA) has also been used interchangeably with MOSA by U.S. DOD.

- Design the open system architecture using open interfaces. Implement the open interfaces using open standards for connecting HW-to-HW, and SW-to-SW.

- The satellite system design shall accommodate growth and provide open interface standards to allow future reconfiguration and addition of new capabilities without large-scale redesign of the system.

- Develop a capability roadmap for the system covering the life of the system following the completion of the rapid prototyping contract phase.

- Address "Commercial Off the Shelf/Non-Developmental Item and Open System Software Licenses," including Open Source Software, Verification of Open Architecture, Modular Open Systems Approach Metrics to be Reported, Modular Open System Approach Analysis Report.

- Generate Open System Management Plan (OSMP) to capture all MOSA activities, technology roadmaps; Define and track MOSA metrics; Update roadmap. Following is a list of MOSA metrics that should be used to demonstrate an open satellite system:

    - Percentage (%) Open Key Interfaces $= \frac{\text{Number Open Key Interfaces}}{\text{Total Number Key interfaces}}$

    - Number and location of private extensions on open interfaces;

    - Contractor use of company private extensions on open standard middleware;

    - Open Software Design Tool Kits/Component Design Tool Kits (OSDTK/CDTK) will be provided with a minimum of Government Purpose Right (GPR); Minimal license fees may apply for COTS items;

    - Percentage of Chief Engineers, IPT Leads and program team members on architecture, software, logistics and Test & Evaluation trained in Open Systems Architecture and the MOSA tools;

    - Future Competition Strategy included in the OA Business plan within the OSMP;

    - MOSA (or OSA) requirements flowed down to sub-tier suppliers and recorded in IBM Rational® DOORS® requirements database or an MBSE digital model.

    - Design a system that consists of hierarchical collections of software, hardware, and firmware Configuration Items (CI's). Document in the MOSA Analysis Report its modularization choices for the system design and any tradeoffs performed in accordance with the OA verification plan.

    - Document any processes or applications necessary to support MOSA in the MOSA Analysis Report.

The above U.S. DOD's guidance encourages the satellite system designers to consider the above MOSA items in the design and build of the modular and open satellite Bus and mission payload for future space systems. The following section presents a proposal for assisting the satellite system designers to implement these MOSA items along with assessment tools provided by U.S. DOD.

4.4 MOSA implementation and assessment tools

It is observed that the U.S. DOD, U.S. civilian agencies (e.g., NASA, NOAA, etc) and U.S. satellite manufacturers/suppliers (e.g., LM, Boeing, Northrop Grumman (NG), Raytheon, L3, etc) are investigating approaches for the modular and open design and build of satellite Busses and mission PL's using MOSA modular and open design principles. **Figure 9** proposes an approach to design and build of future modular and open satellite Busses and mission PLs, and allowing the satellite buyers to: (i) Buy the satellite Bus (see Path A of the figure) and mission PL (see Path B) from different satellite manufacturers/suppliers, (ii) Have an option to choose a third satellite vendor to integrate the satellite Bus and mission PL (see Path C).

The proposed MOSA implementation approach shown in **Figure 4** consists of six basic steps that are incorporated into three execution paths, namely, Path A, Path B and Path C:

- Path A is for the satellite Bus manufacturer/supplier. This path has three basic steps:
    - Step I-A: Develop Modular satellite Bus Architecture (MoBA). The MoBA subsystem components are described in Sections 1 and 2 (see **Figure 3**).
    - Step II-A: Designate KOSS's and select open standards for all internal satellite Bus subsystem components. Open interface standards selection and designation of KOSS are discussed in Sections 3 and 4.

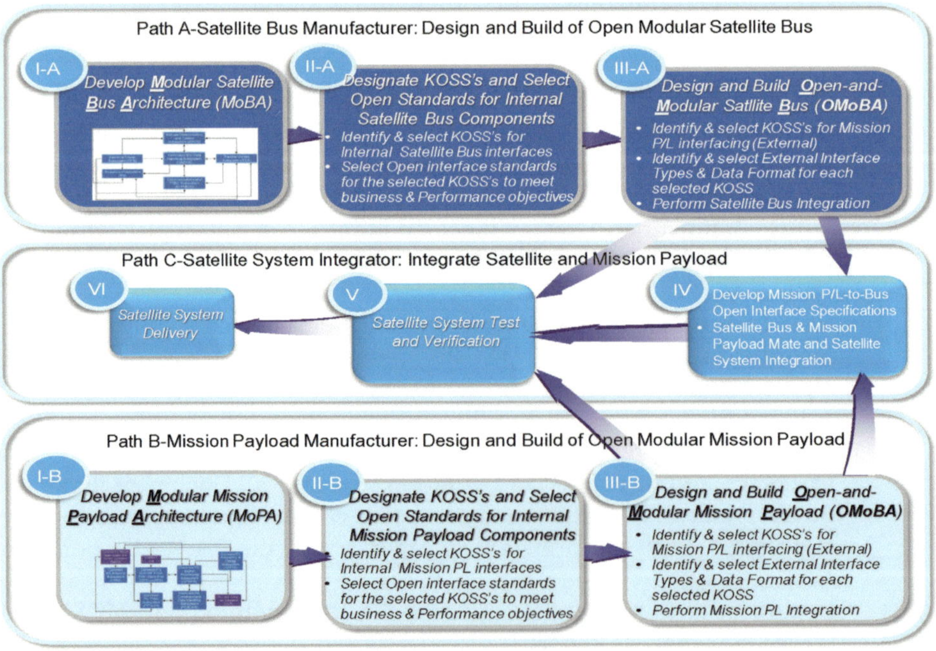

Figure 9.
Proposed implementation approach for design and build of satellite Systems allowing buyers to acquire satellite bus and Mission PL independently.

- Step III-A: Design and build Open Modular satellite Bus System (OMoBS). This step is achieved by identifying all potential KOSS's from the satellite Bus to any mission PL's, i.e., the selected KOSS's should be independent of mission types. The satellite Bus manufacturer is responsible for integrating all Bus components and have the satellite Bus ready for sale.

- Path B is for the mission PL manufacturer/supplier. This path also has three basic steps that are similar to Path A:

    - Step I-B: Develop Modular Mission PL Architecture (MoPA). The MoPA subsystem components are also described in Sections 1 and 2 (see Figure 4).

    - Step II-B: Designate KOSS's and select open standards for all internal Mission PL subsystem components. Open interface standards selection and designation of KOSS for mission PL are also discussed in Sections 3 and 4.

    - Step III-B: Design and build Open Modular Mission PL System (OMoPS). This step is achieved by identifying all potential KOSS's from the any mission PL's to satellite Bus, i.e., the selected KOSS's should be independent of mission types. The mission PL manufacturer is responsible for integrating all mission PL components and have the PL ready for sale.

- Path C is for the satellite system integrator. This path has additional three new steps:

    - Step IV: The system integrator works with satellite Bus and mission PL manufacturers to develop a satellite system interface specification specifying all "open" and "close" interfaces between the mission PL-and-satellite Bus. All open interfaces between the mission PL-and-satellite Bus shall be selected to meet the business and performance objectives approved by the buyer. The system integrator performs satellite Bus and mission PL integration using the approved interface specification.

    - Step V: System integrator performs system test and verification subject to buyer's approval.

    - Step VI: System integrator delivers the satellite system to the buyer.

DOD has also developed MOSA tools to assist MOSA implementation and assessment of military satellite Bus and mission PL "Openness". These tools can also be used for civilian and commercial applications. The DOD tools include MOSA Program Assessment and Rating Tool (PART), Open Architecture Assessment Tool (OAAT), and Key Open SubSystem (KOSS) Tool:

- MOSA PART[3]: It is being used by DOD as the standard MOSA program assessment and rating tool for DOD space system programs.

[3] PART can be found from: https://www.dau.mil/cop/mosa/Lists/Tools/DispForm.aspx? ID=2&ContentTypeId =0x01002BC08FCA204040449CF11CB472BEEE1800AA6D1BC9926604469A02DDB936F94D1F

- MOSA OAAT[4]: Assist U.S. Navy program managers in assessing the "openness" of their programs. It aligns to the Open Architecture Assessment Model (OAAM) as approved by Assistant Secretary of The Navy (ASN) for Research, Development and Acquisition (RDA), which serves as the Navy Acquisition Executive. Other DOD agencies have also been using OAAT since the tool can provide a reproducible and objective method of conducting program assessments.

- MOSA KOSS Tool[5]: One of the key MOSA principles is the Business Principle number 4, namely, Designate Key Interfaces (see Figure 8, B4). The identification of KOSS's is an important task in realizing open systems. This MOSA principle requires the system designers to compromise between cost and performance by selecting a set of KOSS's with their associated interfaces that can be assigned widely used open standards allowing for easy and affordable update and frequent refresh. MOSA KOSS tool provides guidance for KOSS's identification and selection. The tool makes use of system capability road map, system requirements and Subject Matter Expert (SME), program's sponsor and warfighter knowledge to identify the system/subsystem components expected to have a high volatility over the system life cycle. The tool specifies the key interfaces as those either side of volatile components. The tool will help the satellite system designer to identify and rank KOSS's components that will meet both programmatic and technical requirements.

5. Future resilient and robust satellite system architectures

This section demonstrates how to use Steps II-A and II-B of the proposed MOSA implementation approach presented in Section 4.4 for the design and build of future resilient and robust satellite systems. Subsections 5.1 and 5.2 present potential modular-and-open satellite Bus and mission PL architectures, respectively.

5.1 A potential modular open satellite bus architecture solution

To demonstrate how to transition the notional modular satellite Bus system architecture presented in **Figure 3** to a modular-and-open satellite Bus architecture, this subsection provides an example for the transition of three modular Bus subsystems, namely, BC&DH (Bus Subsystem 3), BTT&C (Bus Subsystem 4) and BEPS (Bus Subsystem 5). These modular Bus subsystems are decomposed to subsystem component-level and analyzed for consideration as potential KOSS's for open interface standardization. **Table 1** summarizes the decomposition and analysis results for these three satellite Bus subsystems.

In practice, the preliminary KOSS analysis results shown in **Table 1** should be finalized by the system designer using DOD KOSS tool discussed in Section 4.4. As shown in **Table 1**, standardizing the BC&DH data interfaces will probably provide the biggest return on investment since the BC&DH subsystem interfaces with each onboard system. Incorporation of the timing interface along with the data interface will minimize the amount of connections, thus reducing overall system mass. Any

---

[4] OAAT from: https://www.dau.mil/cop/mosa/Lists/Tools/DispForm.aspx?ID=1&ContentTypeId=0x01002 BC08FCA204040449CF11CB472BEEE1800AA6D1BC9926604469A02DDB936F94D1F

[5] KOSS from: https://acc.dau.mil/adl/enUS/317012/file/46502/KOSS%20Overview_FINAL_5Aug09.pdf.

Modular satellite Bus subsystem component	Modular satellite Bus subsystem and component description	Recommendation for open interface Standardization (potential KOSS)
BC&DHS Component No.	Bus Command & Data Handing Subsystem (C&DHS)	
BC&DHS-1	Command Authentication Processing Unit (Sync Word Frame Lock, Unparsed Command)	Recommend for Interface standardization
BC&DHS-2	System Timing Unit	
BC&DHS-3	Fault Management Processing Unit (Execute Stored CMD Sequence, Monitor System Health)	Not recommended for interface standardization due to many variations between systems
BC&DHS-4	Bus Resource Management Processing Unit (Managing Internal and External Bus Data)	Recommend for open interface standardization
BC&DHS-5	Memory Storage Unit	
BC&DHS-6	Spacecraft Control Processor	
BC&DHS-7	Bus Telemetry Conditioning Processor	Not required; software driven functions. Should be considered in software interface analysis.
BC&DHS-8	Bus Cyber Security Unit	Recommend for open interface standardization
BTT&CS Component No.	Bus Tracking-Telemetry & Command Subsystem (TT&CS)	
BTT&CS-1	TT&C Waveforms/MODEM	Recommend for open interface standardization
BTT&CS-2	TT&C Antenna Assembly for S-Band/L-Band	Not recommended for interface standardization.
BTT&CS-3	TT&C RF Front-End and Back-End Assembly	Recommend for open interface standardization
BTT&CS-4	Unified S-Band (USB) RX/TX Assembly	
BTT&CS-5	SGLS S-Band RX/TX Assembly	
BTT&CS-6	SGLS Base Band Signal Processing (USB Mode1, 2)	Recommend for open interface standardization
BTT&CS-7	In Band TT&C Processor Located at Private Station	Not recommended for interface standardization due to many variations between systems
BTT&CS-8	Power Controller Assembly	Recommend for open interface standardization
BEPS Component No.	Bus Electrical Power Subsystem (EPS)	
BEPS-1	Solar Array (SA)	Recommend for open Interface standardization.
BEPS-2	Battery Assembly (BA)	Not recommended for interface standardization; Battery size will vary depending on the mission profile. Additional batteries could potentially require customized interfaces to tie all batteries to power bus.

Modular satellite Bus subsystem component	Modular satellite Bus subsystem and component description	Recommendation for open interface Standardization (potential KOSS)
BEPS-3	Solar Array Drive Assembly (SADA)	Recommend for open Interface standardization.
BEPS-4	Transient Filter Unit (TFU)	Not recommended for interface standardization
BEPS-5	Bus Power Regulation Unit (BPRU)	Recommend for open Interface standardization.
BEPS-6	Fuse Box Assembly (FBA)	
BEPS-7	Pyro Relay Assembly (PRA)	

Table 1.
Satellite bus subsystems decomposition and potential KOSS.

interfaces that require a significant amount of analysis or Non-Recurring Engineering (NRE) hours is not a good candidate for standardization. The fault management processing interface is in this category, and it is not recommended for standardization.

5.2 A potential modular open Mission payload architecture solution

This subsection provides an example for the transition of the notional modular mission PL architecture presented in **Figure 4** to a modular-and-open mission PL architecture. **Table 2** summarizes the decomposition and KOSS analysis results for four mission PL subsystems, including PAS (PL Subsystem 1), CPCom-RFS (PL Subsystem 2), PDPS (PL Subsystem 3) and PFTS (PL Subsystem 11).

The mission PL digital processing system is not recommended for interface standardization due to many variations between systems and subsystems. Multi-RF Wideband RX Up/Down Converters and Tunable IF Down Converters require a significant amount of analysis or NRE hours and are also not a good candidate for standardization. Again, DOD KOSS tool should be used to finalize the KOSS analysis results presented here for actual design and build of the satellite systems.

6. Conclusion

The chapter provides an overview of existing modular satellite Bus and mission PL architectures and associated standards for communication data Busses. The chapter defines open and close interfaces along with industry approved popular standards and discusses the interface design challenges. Moreover, the chapter provides an overview of MOSA and related DOD guidance and assessment tools to address the interface design challenges. Examples for the design and build of modular-and-open satellite Bus and mission PL architectures are also presented. The intent of this chapter is to provide an innovative approach for the satellite system designer to design and build of the next generation satellite achieving a balance between business and technical objectives that make a business sense for both the satellite manufacturers and buyers in terms of lower system acquisition and sustainment costs over its life cycle. The MOSA implementation approach presented here allows the satellite manufacturers to build the satellite Bus and mission PL separately for more production, flexibility, and market competition. Concurrently, the approach also allows the satellite buyers to buy satellite Bus at high volume with reduced unit costs and less schedule risk. Another benefit for the

Modular mission PL subsystem component	Modular mission PL subsystem and component description	Recommendation for open interface standardization (potential KOSS)
PAS Component No.	PL Antenna Subsystem (PAS)	
PAS-1	PL RF Antenna Configuration (PRAC)	Not recommended for interface standardization due to many variations between systems
PAS-2	Beam Forming Unit (BU)	
PAS-3	Antenna Controller (AC)	Recommend for Open Interface standardization
PCom-RFS Component No.	PL Com RF Front-End/Back-End Subsystem (CPCom-RFS)	
PCom-RFS-1	PL LNA Component	Not recommended for interface standardization due to many variations between systems/ subsystems
PCom-RFS-2	PL HPA Component	
PCom-RFS-3	Multi-RF Wideband Receiver (RX)	
PcCm-RFS-4	PL Up/Down Converters	Recommend for Open Interface standardization
PCom-RFS-5	Tunable IF Down Converters	
PDPS Component No.	PL Digital Processing Subsystem (PDPS)	
PDPS-1	PL ADC/DAC	Not recommended for interface standardization due to many variations between systems
PDPS-2	FPGA Processor	
PDPS-3	PL MOD and DEMO (Optional)	
PDPS-4	Digital Network Switch (Optional)	
PDPS-5	PL System Controller	
PTFS Component No.	P/L Frequency and Timing Subsystem (PFTS)	
PFTS-1	Atomic Clock Unit (ACU)	Not recommended for interface standardization; ACU and CM&CU will vary depending on mission type and mission requirements
PFTS-2	Clock Monitoring & Control Unit (CM&CU)	
PFTS-3	Frequency Generation & Up conversion Unit	Recommended for Interface standardization.
PFTS-4	Timing Variation and Frequency Stability	Not recommended for interface standardization

Table 2.
Mission payload subsystems decomposition and potential KOSS.

satellite buyer is the adaptability of changing the requirements on the mission PL without impacting the satellite Bus.

Acknowledgements

Although the preparation of this work was not funded by The Aerospace Corporation, but the author acknowledges the work presented in this chapter was based on his knowledges accumulated over the years from many space programs at The Aerospace Corporation, Raytheon and Jet Propulsion Laboratory. In addition, the author would like to express his appreciation to Aerospace's manager, Ms. Navneet Mezcciani, for her professional support.

Conflict of interest

The preparation of this chapter was not funded by the Aerospace Corporation, and it was done by the authors using his own time and resources, thus it does not represent the Aerospace Corporation's view on the DOD guidance for MOSA implementation and the proposed system architecture solutions.

Notes/Thanks/Other declarations

The author wishes to thank his wife, Thu-Hang Nguyen, for her enormous patience and boundless support during the preparation of this chapter.

Author details

Tien M. Nguyen[1,2]

1 California State University, Fullerton, USA

2 The Aerospace Corporation, El Segundo, USA

*Address all correspondence to: tmnguyen57@fullerton.edu; tnguyen57@aol.com

References

[1] U.S. DOD Open Systems Architecture Contract Guidebook for Program Managers, v.1.1, June 2013. Available from: https://www.dau.edu/cop/pm/layouts/15/WopiFrame.aspx?sourcedoc=/cop/pm/DAU%20Sponsored%20Documents/DoD%20Open%20System%20Architecture%20(OSA)%20Contract%20Guidebook%20for%20Program%20Managers-version%201.1-%20June%20%202013.pdf&action=default&DefaultItemOpen=1

[2] Zimmerman P. Modular open Systems approaches (MOSA) panel on standards. In: Office of the Deputy Assistant Secretary of Defense on Systems Engrg, Defense Standardization Program Workshop. 12 July 2018. pp. 1-7

[3] U.S. Department of Navy, Naval Open Architecture Contract Guidebook for Program Managers, V.2.0. 30 June 2010. Available from: http://www.dtic.mil/dtic/tr/fulltext/u2/a550060.pdf

[4] Acquiring and Enforcing the Government's Rights in Technical Data and Computer Software Under Department of Defense Contracts: A Practical Handbook for Acquisition Professionals. 7th ed. Los Angeles, CA: Office of the Staff Judge Advocate, Space and Missile Systems Center (SMC), U.S. Air Force Base; August 2015. Available from: https://www.acq.osd.mil/dpap/cpic/cp/docs/Technical_Data_and_Computer_Software_Rights_Handbook_7th_Edition.pdf

[5] Revillon P. Fundementals and dynamics of the satellite communications business. In: 2016 EuroConsult for Inmarsat Capital Markets Day. Available from: www.euroconsult-ec.com

[6] Davis LA, Filip L. How long does it take to develop and launch government satellite systems? In: Aerospace Report No. ATR-2015-00535. El Segundo, California: The Aerospace Corporation; 2015

[7] Boeing. Boeing Satellites. 2015. Available from: http://www.boeing.com/space/boeing-satellite-family/

[8] BAE Systems. Processors. 2015. Available from: http://www.baesystems.com/down load/BAES_161071/processors

[9] GAO. Space Acquisitions – Space Based Infared Systems Could Benefit from Technology Insertion Planning. 2015. Available from: http://www.gao.gov/assets/670/669454.pdf

[10] GPS. Global Positioning System Constellation Replenishment. 2015. Available from: http://www.gps.gov/congress/reports/2015/2015-02-AF-report-GPS-constellation-replenishment.pdf

[11] Martin L. A2100. 2015. Available from: http://www.lockheedmartin.com/content/dam/lockheed/data/space/documents/a2100/A2100brochure.pdf

[12] L Martin. Advanced Extremely High Frequency (AEHF). 2015. Available from: http://lockheedmartin.com/us/products/advanced-extremely-high-frequency–aehf.html

[13] Martin L. Global Positioning System (GPS). 2015. Available from: http://www.lockheedmartin.com/us/products/gps.html

[14] Loral. 1300 Series Satellite Platform. 2015. Available from: http://www.sslmda.com/html/products/1300.html

[15] National Aeronautical and Space Administration (NASA). Tracking and Data Relay Satellite (TDRS) Third Generation Capabilities. 2013. Available from: http://www.nasa.gov/directorates/

heo/scan/services/networks/txt_tdrs_gen3.html

[16] Collins R. Momentum and Reaction Wheels. Available from: https://www.rockwellcollins.com/Data/Products/Space_Components/Satellite_Stabilization_Wheels/RSI_12_Momentum_and_Reaction_Wheels.aspx

[17] Gwaltney DA, Briscoe JM. Comparison of Communication Architectures for Spacecraft Modular Avionics Systems. NASA/TM—2006–214431. June 2006. NASA Center for AeroSpace Information

[18] SpaceWire—Links, Nodes, Routers, and Networks, ECSS–E–50–12A. 24 January 2003. European Cooperation for Space Standardization

[19] NASA Common Instrument Interface Project. Hosted Payload Guidelines Document. Document Number: CII-CI-0001, Version Rev. A., 4 November 2013. Earth System Science Pathfinder Program Office

[20] NASA Common Instrument Interface Project. Hosted Payload Guide for Proposers. Document Number: HPIG0001, Version: Initial. 22 March 2018

[21] Space and Missile Systems Center. Electromagnetic Compatibility (EMC) Requirements for Space Equipment and Systems, SMC Standard SMC-S-008. 13 June 2008. Air Force Space Command

[22] NASA. Electrical Grounding Architecture for Unmanned Spacecraft, NASA HDBK-4001. 17 February 1998. NASA Technical Handbook

[23] Plancke P. Spacecraft Digital Busses and Interfaces Selection and Standardization Status. 19–20 September 2001. On-Board Payload Data Processing Workshop ESTEC. Available from: ftp.estec.esa.nl/pub/ws/OBCDS_Workshop/Proceedings.htm

[24] Consultative Committee for Space Data Systems (CCSDS). Recommendations for Space Data Systems, Packet Telemetry, CCSDS 102.0-B-5, Blue Book. 2000. Available from: https://public.ccsds.org/Pubs/102x0b5s.pdf#search=packet%20telecommand

[25] Consultative Committee for Space Data Systems (CCSDS). Recommendations for Space Data Systems, Telecommand – Part 1: Channel Service, CCSDS 201.0-B-3, Blue Book. 2000. Available from: https://public.ccsds.org/Pubs/201x0b3s.pdf#search=packet%20telecommand

[26] Space-Ground Link Subsystem (SGLS). Ground Station, System Analysis Summary Report, WDL-TR3227-1. Vol. 1. 15 November 1968. Philco-Ford Corporation. Available from: https://apps.dtic.mil/dtic/tr/fulltext/u2/853122.pdf

[27] Hearn M. MIL-STD-1553B: The Past and Future Data Bus. 2019. High Frequency Electronics Magazine. Available from: https://www.highfrequencyelectronics.com/index.php?option=com_content&view=article&id=1955:mil-std-1553b-the-past-and-future-data-bus&catid=161&Itemid=189

[28] Department of Defense Interface Standards for Digital Time Division Command/Response Multiplex Data Bus, MIL-STD-1553B. 21 September 1975, Department of Defense, Washington D.C. 20360

[29] AeroFlex, SpaceWire 101, Webex Seminar. 15 February 2006. Available from: www.aeroflex.com/spacewire

[30] ESA Publications Division. SpaceWire Standard Document

ECSS-E-ST-50-12C. The Netherlands; 31 July 2008

[31] Los Angeles Air Force Base (LAAFB). Space and Missile Systems Center (SMC), Common Payload Interface Specification Project (CoPaIS), SMC18-069. 5 April 2018. Available from: https://govtribe.com/opportunity/federal-contract-opportunity/common-payload-interface-specification-project-copais-smc18069

Chapter 6

System Designs of Microsatellites: A Review of Two Schools of Thoughts

Triharjanto Robertus

Abstract

Microsatellite has been considered as disruptive technologies in satellite engineering. Its development cost and time provide advantages for new kind of Earth observations, telecommunications, and science missions. The increasing trend of microsatellite launches and operations means that the approach was so successful that it could create funding sustainability. Major contributing factors of its success were due to the system design of the microsatellites. This chapter discusses two microsatellite system design approaches, namely Technical University of Berlin heritage and University of Surrey heritage. Both Universities provide approaches for system design and build of microsatellite systems. The design approaches are being compared along with lessons learned. The choices of microsatellites to be compared in this chapter will be those that are manufactured about the same time such that the technology compared is mostly the same and flown in-orbit. The chapter shows that the differences between the two system design approaches are on the choice of main computer and associated link configuration and in the attitude control modes. Another major different is in the satellites' structure design. For some satellite's components, incoming technologies have made the design choices from the two schools of thoughts converged.

Keywords: satellite design, system design, microsatellites, TU Berlin, University of Surrey

1. Introduction

Microsatellite has typical weight between 20 and 170 kg at launch as auxiliary payload. It is initially made as technology experiment and education tools by universities. Nowadays, microsatellite becomes a common space platform for commercials and emerging space nations. The commercial mission is typically Earth observation, data collecting platform (text-based communication), including ships and aircraft tracking. Studies done by Swartout [1] show that between 2009 and 2012, about 8–12 satellites with mass above 50 kg as auxiliary payload were launched yearly. The data also show that the trend seems to be steady. Bunchen and De Pasquale [2] noted that 105 satellites with mass of 11–50 kg were launched between 2000 and 2013.

Surrey Space Technology Limited (SSTL), a subsidiary company under University of Surrey, is one of the companies that initiated the use microsatellite technology as commercial Earth observation satellite platform. It built a constellation of five satellites named Disaster Monitoring Constellation (DMC) in 2003, with payload of 3-band multispectral imager of 30-m resolution, which was intended for wide-swath land coverage imaging. After the first constellations decommissioned, it built the second generation with better resolution (20 m). The first launch of DMC-2 constellation was done in 2009 [3].

Since 2013, Skybox/Skysat has deployed 15 satellites that carry 1-m panchromatic imager and 2-m 4-band multispectral imager [4]. Unlike DMC, which mission objectives are to observe wide areas with nadir pointing scanning mode, it aims to provide frequent repeat very high resolution images using massive numbers of highly maneuverable satellites. Another commercial Earth observation microsatellite constellation mission is prepared by Axelspace. The company planned to have 50 satellites launched starting 2017. The satellite carries imager with 2.5-m panchromatic and 5-m multispectral [5, 6]. **Figure 1** shows the configurations of the Skybox and Grus satellites, which show that Skybox uses single lens and parabolic data downlink antenna, while Grus uses two lenses and horn-type data downlink antenna.

In addition to Earth observation missions, microsatellite constellation also being built for Low Earth Orbit (LEO) telecommunication mission. OneWeb and Telesat are two companies that will launch hundreds of microsatellites in coming years [7, 8].

The use of microsatellites for commercial purposes means that the technology is mature enough to ensure good return-of-investment. One of the major aspects that contribute to the success of microsatellite technology is its system design. Therefore, the objective of this chapter is to provide insight into microsatellite system design. The chapter addresses the question related to limitation in weight and size, and how the satellite designer manages to meet the mission requirements.

Out of many microsatellites developers, two system designs of microsatellites, namely Technical University (TU) Berlin heritage and University of Surrey heritage, are selected for comparison in this chapter, due to their very different design approaches. To be comparable, the choices of microsatellite system to be compared are the ones that manufactured about the same time, so that the technology available is mostly the same. The microsatellites also have to have in-orbit experience, so its design success can be measured. Data mining resulted that the satellite operation year chosen is between 1999 and to date. For TU Berlin system, the choices are DLR-TUBSAT, MAROC-TUBSAT, Indonesian LAPAN-TUBSAT, LAPAN-ORARI, and LAPAN-IPB. Meanwhile, for University of Surrey system, the choices are Korean KITSAT-3, STSat-1 and STSat-3, as well as Turkish BILSAT-1 and RASAT.

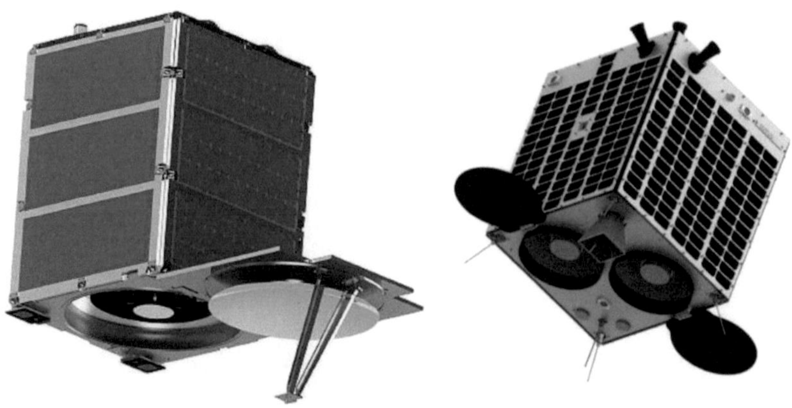

Figure 1.
Google Skybox satellite and Axelspace's Gruz satellite design.

This chapter is divided into five sections, with the first section introducing the background and objectives of the chapter. The second section explains how the satellite design samples for the University of Surrey heritage were selected, and what satellite design parameters were used in the comparison. Section 3 displays the satellite design parameters for TU Berlin heritage. Section 4 provides analysis from the comparison of the two-design heritage, in term of parameters noted in the previous two sections. Section 5 summarizes the analysis and provides recommendation for further studies regarding the subject.

2. University of Surrey heritages

University of Surrey is known as one of the pioneers in the design and build of microsatellite in the 1990s. It started launching microsatellite in 1991 with amateur

	KITSAT-3	BILSAT-1
Operation	1999–2003	2003–2006
Bus		
Solar panel	3 GaAs (2 deployable) @ 50 × 85 cm, (150 W)	4 GaAs @ 60 × 60 cm (58 W)
Battery	NiCd; 10 V; 8 Ah	NiCd; 28 V; 4 Ah
Reaction wheel/Gyro	3 + 1 Teldix DR01/FO laser	4 SSTL/MEMS
Thruster	—	Pressurized gas + resistojet
Star sensor	1	2 Altair
Sun sensor	2 axis	4 × 2 axis
Horizon sensor	2 axis	
Magnetotorquer/meter	3-axis air coils/3-axis fluxgate	3-axis air coils/2 × 3-axis fluxgate
Telemetry, Tracking, and Commanding (TTC)	VHF uplink; UHF downlink	S-band
Data TX	S-band 3.3 Mbps	S-band 8 Mbps
Main computer/link config.	2 × microprocessor/CAN	2 × microprocessor/CAN
Attitude control computer	1	1
Payload data handling	Microprocessor based	FPGA based
GPS	—	SSTL SGR
Payload		
	3-band imager w/ 570-mm lens	2 × 3-band imager w/150-mm lens
	Radiation dose sensor	Pan imager w/300 mm lens
	High energy particle sensor	Store and forward communications
	Scientific class magnetometer	8-band low resolution imager
		CMG
Size (cm)	50 × 60 × 85	60 × 60 × 60
Mass (kg)	110	130

Table 1.
Sample for the University of Surrey microsatellite system design.

radio missions. To simplify the satellite design, the first microsatellite generation has passive attitude control system, that is, using gravity gradient telescopic boom. The university provided microsatellite development and building capabilities to many emerging space countries, including Thailand, Malaysia, South Korea, Algiers, Turkey, and Nigeria. At the time, such countries started to use remote sensing satellites, mostly from the United States and European, for various land-based applications. Therefore, they required remote sensing payloads to include in their satellite missions. Such mission elevates the design requirements to active attitude control system and higher data rate downlink system.

Thailand's Mahanakorn University collaborated with the University of Surrey to jointly develop TMSat that was launched in 1998 [9]. TMSat focuses on remote

	STSAT-1	STSAT-3	RASAT
	2003–2008	2013–2015	2011–2017
Bus			
Solar panel	3 GaAs (2 deployable); 160 W	3 GaAs (2 deployable); 275 W	4 GaAs; 52 W
Battery	NiCd; 14 V; 12 Ah	Li-ion; 20 V; 20 Ah	Li-ion; 28 V; 9 Ah
Reaction wheel/Gyro	4 /FO laser	4 /FO laser	4 /MEMS
Thruster	—	Hall thrust	—
Star sensor	1	2 SaTReC	1
Sun sensor	4 panels +2 cell	Coarse and fine	4 analog
Horizon sensor	—	—	—
Magnetotorquer/meter	3-axis/3-axis fluxgate	3-axis/3-axis	3-axis/2 × 3-axis fluxgate
TTC	S-band	S-band	S-band (primary) and UHF/VHF (emergency)
Data TX	X-band 3.2 Mbps	X-band 10 Mbps	X-band 100 Mbps
Main computer/link config.	Microprocessor/CAN	Leon2-FT (triple redundancy)/CAN and space wire	2 × microprocessor/ CAN and space wire
Attitude control computer	1	1 AIU (attitude interface unit)	1
Payload data handling	FPGA based	FPGA based	FPGA based
GPS	1	1	1
Payload			
	Far UV imaging spectrograph	2× Multiband IR imagers	Pan imager w/840 mm lens
	Space physic sensor	Spectrometer	3-band imager w/420 mm lens
	Data collection system		
Size (cm)	66 × 55 × 83	102 × 103 × 88	70 × 70 × 55.4
Mass (kg)	106	175	95

Table 2.
Sample for the University of Surrey microsatellite heritage system design.

sensing and amateur radio mission. Since Thailand did not continue building its subsequent satellites, TMSat is not selected as satellite design heritage sample in this chapter.

Singapore's Nanyang Technology University (NTU) collaborated with the University of Surrey to jointly develop satellite subsystem for UoSAT-12. However, the satellite is not a microclass and therefore is not selected as a sample for the University of Surrey's satellite system design in this chapter. The satellite subsystem from NTU is a communication payload with S-band downlink and L-band uplink, which provides the Internet protocol communication operating at 1 Mbps. Since the experience with the University of Surrey only in subsystem design and development, the subsequent NTU satellite, that is, XSAT, is also not considered as the University of Surrey heritage satellite [10, 11].

South Korean experience with the University of Surrey satellite design is when Satellite Technology Research Center (SaTReC), an institution under Korea Advanced Institute of Science and Technology (KAIST), jointly built KITSAT-1 and KITSAT-2 and launched it in 1992 and 1993. Both satellites have store-forward communication amateur payload and low-resolution imagers. Since the KITSAT-1 and KITSAT-2 development time does not match with other microsatellite design sample, only the design of KITSAT-3 is used in this chapter. SaTReC then developed STSAT series as its second generation microsatellites. Since STSAT-2 experienced launch failure, only STSAT-1 and STSAT-3 are selected as satellite design samples [12–15].

Turkey's experience with the University of Surrey satellite design is when its space research institute, TUBITAK-UZAY (previously named BILTEN TUBITAK-ODTU), jointly developed BILSAT-1. The satellite was part of DMC-1 constellation [16–19]. After BILSAT-1, the institute then built its second generation microsatellite, RASAT. Therefore, both microsatellites are used as sample for the University of Surrey design heritage [20–23].

Fifteen satellite bus design parameters are selected for the comparison, including 14 mechatronics component parameters in the satellites' design. For the University of Surrey satellite heritage, the parameters are tabulated in **Tables 1** and **2**. Structure design from four of the five microsatellites is shown in **Figures 2** and **3**. Payload parameters also noted in **Tables 1** and **2** to explain the similarity (or differences) in the mission requirements and their impacts to satellite bus parameters. The weight and dimensions are, in additional of drawings, noted in to explain the satellite

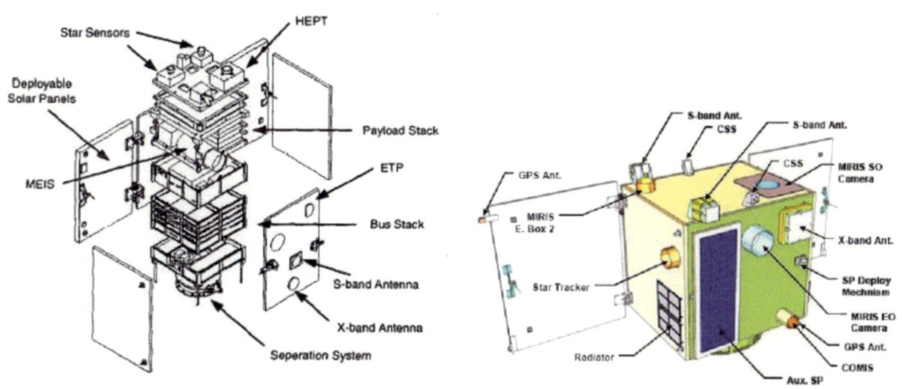

Figure 2.
Mechanical design of KITSAT-3 and STSAT-3.

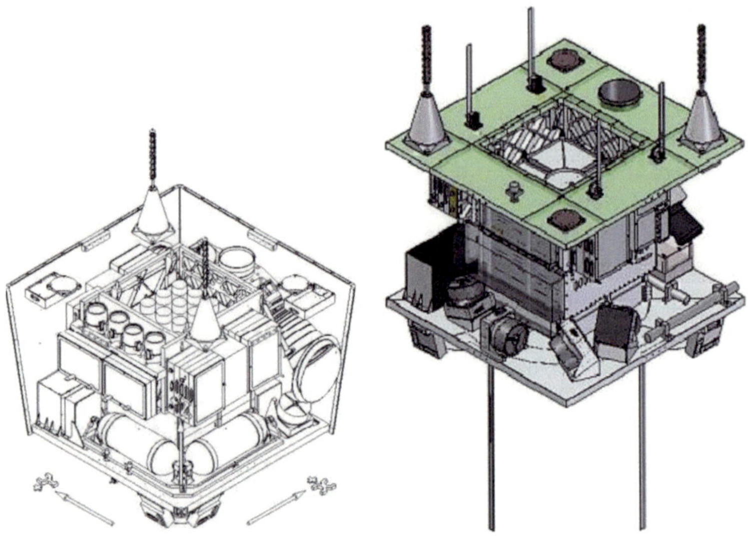

Figure 3.
Mechanical design of BILSAT-1 and RASAT.

structure design aspects. The satellite operation years are noted in the tables to show the context of available technology.

As shown in **Figures 2** and **3**, the University of Surrey heritage satellites use electronic trays for its satellite bus electronics. The aluminum trays also function as load bearing structure, so that the rest of the satellites components, such as reaction wheels and attitude sensors, can be laid out around them. After all components integrated, the solar panels and/or other outside panels that are made of lighter materials can be used to cover the satellites.

3. Technical University of Berlin heritages

Technical University (TU) of Berlin had launched six microsatellites between 1991 and 2007. During such time, the university had provided microsatellite development capacity building to Morocco and Indonesia. However, only Indonesia (Satellite Technology Center) had developed its second generation of microsatellites. **Tables 3** and **4** provide samples of microsatellite systems used for the comparison considering the development and operation time of the satellites. The microsatellite parameters from the TU Berlin heritage shown in **Table 3** are from DLR-TUBSAT and MAROC-TUBSAT, and in **Table 4** are from LAPAN-TUBSAT and two Indonesian built satellites, that is, LAPAN-ORARI and LAPAN-IPB. Additionally, the four satellite structure drawings are presented in **Figures 4** and 5 [24–29] for the comparison of structural design.

The author should describe the key differences among the four structure designs presented in **Figures 4** and 5.

As shown in **Figures 4** and **5**, for the TU Berlin satellite heritage, the components are laid out in boxes. For DLR-TUBSAT and Maroc-TUBSAT, they are modular boxes (ACS, payload, power, etc.). Meanwhile, in LAPAN's satellite series, the boxes are integrated in lower and upper compartments of the same structure. The boxes were made from aluminum plates and therefore function as load bearing structure. The solar panels are directly attached to the outer part of the boxes.

	DLR-TUBSAT	MAROC-TUBSAT
Launch	1999–2007	2001–2006
Bus		
Solar panel	4 Si @32 × 32 cm, (14 W)	4 Si @32 × 32 cm (14 W)
Battery	NiH_2; 10 V; 12 Ah	NiH_2; 10 V; 12 Ah
Reaction wheel/Gyro	3 IRE 203/FO laser	3 + 1 IRE 203/FO laser
Thruster	—	—
Star sensor	—	IRE
Sun sensor	4 panels +1 cell	6 single cell
Horizon sensor	—	—
Magnetotorquer/meter	1 axis coil + 1 rod/–	1 axis/3-axis sensor
TTC	2 UHF w/omni antennas	2 UHF w/omni antennas
Data TX	S-band analog	S-band 256 kbps
Main computer/ link config.	32 bit microcontroller/star	32 bit microcontroller/star
Attitude control computer	—	—
Payload handling	Multiplexer	Recorder
GPS	—	—
Payload		
	B/W video cam. w/16 mm lens	NIR imager w/72 mm lens
	B/W video cam. w/50 mm lens	
	B/W video cam. w/1000 mm lens	
Size (cm)	32 × 32 × 32	32 × 34 × 36
Mass (kg)	45	47

Table 3.
Sample for the Technical University of Berlin microsatellite system design.

4. Analysis

4.1 Power generation and storage

Tables 1 and **2** show that the Korean satellites have employed deployable solar panel (which is also shown in **Figure 1**), since the mission required high power and used direct energy transfer (DET) mode. Such approach is very much different than those used by KITSAT-1 and KITSAT-2, which have body-mounted solar panels. On the other hand, Turkish satellites use body-mounted solar panels and therefore do not have the requirement of one side of the satellite always facing the sun for battery charging.

Tables 3 and **4** show that all TU Berlin heritage use body-mounted solar panels. It uses Si panels for its first three satellites, then opted to higher capacity GaAs panels in LAPAN-ORARI and LAPAN-IPB. Generally, the power budget for the University of Surrey heritage satellites is higher than the TU Berlin heritage, even in the ones with body-mounted solar panels. As shown in **Figure 5**, in LAPAN-IPB, one of the sides has two 46 × 26 cm solar panels. The side is projected to be Sun pointing most of the time.

Battery chosen to be used in the early University of Surrey heritage satellite design is NiCd, while in TU Berlin's satellite design is NiH_2. NiCd batteries require

	LAPAN-TUBSAT	LAPAN-ORARI	LAPAN-IPB
	2007–2013	2015-now	2016-now
Bus			
Solar panel	4 Si @43 × 24 cm, (14 W)	4 GaAs @46 × 26 cm (30 W)	5 GaAs @46 × 26 cm (30 W)
Battery	NiH_2; 14 V; 12 Ah	Li-ion; 16 V; 19.5 Ah	Li-ion; 16 V; 36 Ah
Reaction Wheel/Gyro	3 IRE 203/FO Laser	3 + 1 IRE 303/FO Laser	3 + 1 IRE 303/FO Laser
Thruster	—	—	—
Star sensor	Vectronics (VTS)	VTS, IRE	VTS, LAPAN
Sun sensor	4 panels +2 cells	6 single cells	6 single cells
Horizon sensor	—	—	LAPAN (IR camera based)
Pitch sensor	—	—	LAPAN (CCD based)
Coil/magnetometer	3 axis/–	3 axis/VFMS-51	3 axis/fluxgate scientific class
TTC	2 UHF w/ omni antennas	2 UHF w/ omni antennas	2 UHF w/omni antennas
Data TX	S-band analog	S-band 5 Mbps	X-band 105 Mbps
Main computer/link config.	32 bit microcontroller/ star	32 bit microcontroller/ star	32 bit microcontroller/ star
Attitude control computer	—	—	—
Payload handling	Multiplexer	Digital and analog switcher + recorder	FPGA based
GPS	—	VGPS-51	VGPS-51
Payload			
	Color video cam. w/50 mm lens	Color video cam. w/1000 mm lens	4-band imager w/300 mm lens
	Color video cam. w/1000 mm lens	4 M pix cam. w/1000 mm lens	4 M pix cam. w/1000 mm lens
		AIS (ship monitoring system)	AIS (ship monitoring system)
		APRS (amateur text message)	
		Amateur voice repeater	
Size (cm)	45 × 27.5 × 45	47 × 38 × 50	50 × 57.4 × 42.4
Mass (kg)	54.7	74	115

Table 4.
Sample for the Technical University of Berlin microsatellite heritage system design.

charging controller mechanism ensuring that the battery is completely drained before being charged. This is because partial charging can induce memory effect, which can decrease the battery capacity to its last partial charge state. For NiH_2 batteries, they tend to have large packaging due to its cylindrical shape, as shown in DLR-TUBSAT and LAPAN-TUBSAT drawing (**Figures 4** and **5**), but its charging mechanism is very simple (can do trickle charging). As soon as Li-ion battery technology available, both designs opted out Li-ion battery for its easy handling (no memory effect) and higher power-to-mass ratio.

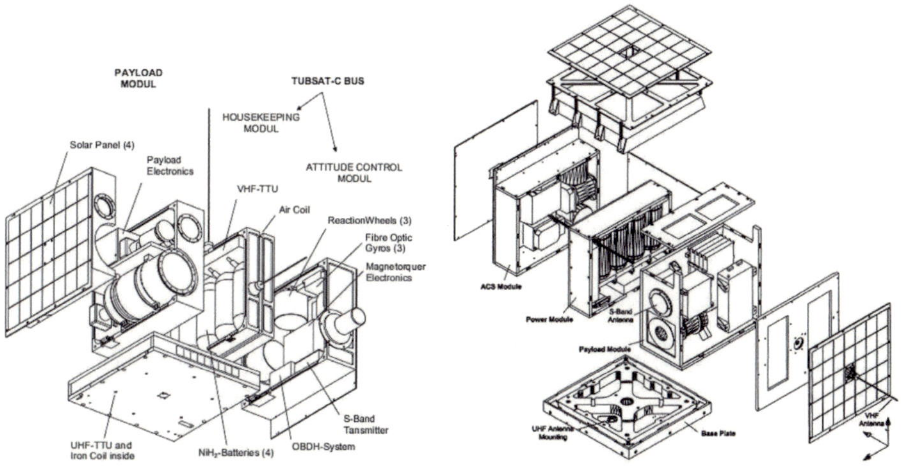

Figure 4.
Mechanical design of DLR-TUBSAT and MAROC-TUBSAT.

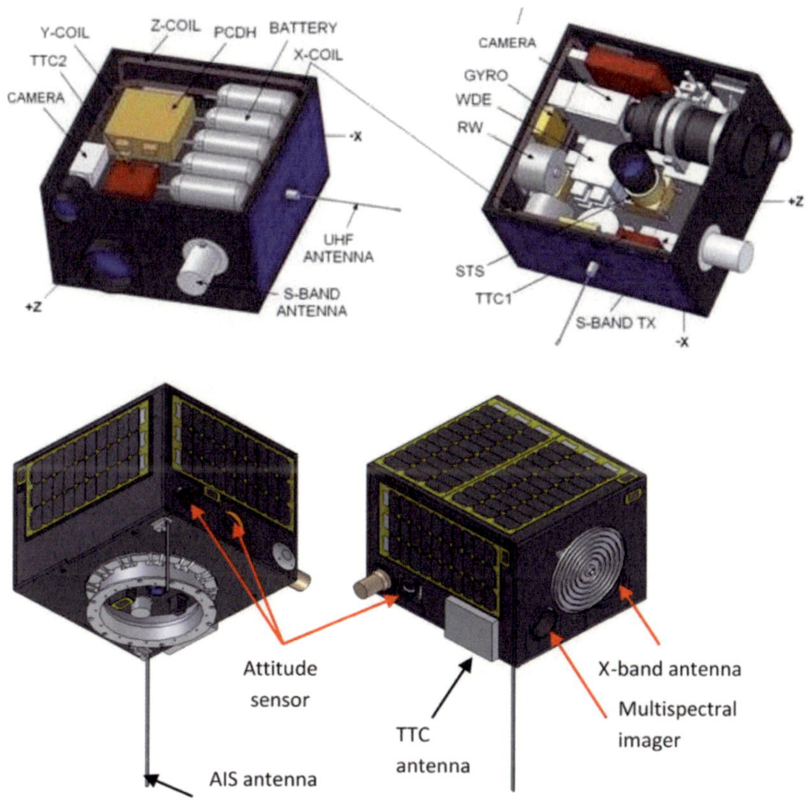

Figure 5.
Mechanical design of LAPAN-TUBSAT and LAPAN-IPB.

4.2 Main computer

On the choice of main computer, the University of Surrey heritage uses microprocessor, such as 32-bit PowerPC 603, while the TU Berlin heritage uses microprocessor, such as 32-bit SH series. Advantage of using microcontroller is having shorter booting time, so that it can recover quickly in the event of latch-up and needs to be restarted. The advantage of microprocessor is its ability to handle more complex and parallel jobs. To anticipate any anomaly in the operation, the use of microprocessor is usually done by using redundancy (i.e., a second processor will take over the operation in the event of anomaly). In the University of Surrey satellite design heritage, the electronic components are connected to main computer with dual line of controller area network (CAN). Meanwhile, the TU Berlin satellite design heritage uses star configuration with dedicated line to each component from the main computer, using RS232 or 422.

4.3 Attitude control subsystem

Tables 1 and **2** show that the University of Surrey satellite design heritage uses separate attitude control computer that integrates attitude sensors, including sun and star sensors with all reaction wheels and gyros. This is done so that the attitude control system can work in closed loop all the time. Such approach is necessary for the microsatellite design with deployable solar panels, such as KITSAT-3, STSAT-1, and STSAT-3 since failure of sun pointing could be disastrous for the satellite. As shown in **Tables 3** and **4**, in the TU Berlin satellite design heritage, none of the satellites have separate attitude control computer. In the design, each reaction wheel-gyro pair directly connected to the main computer, and therefore, closed loop with star and sun sensors can only be done using the main computer resources.

Differences are also found in the attitude control sensor between the University of Surrey design heritage. The Korean microsatellites use fiber-optic gyro, while the Turkish microsatellites use MEMS gyro. Meanwhile, in all TU Berlin microsatellites, fiber-optic gyros are used.

For attitude control actuators, all the selected satellites use reaction wheels and air coils for angular momentum dumping/generation. Figures and data showed that TU Berlin heritage satellites use reaction wheels in 3-axis configuration. For LAPAN-ORARI and LAPAN-IPB satellites, they used redundant wheel at satellite major inertia axis that noted as 3 + 1 as shown in **Table 4**. For the University of Surrey heritage satellites, only KITSAT-3 uses reaction wheels in 3-axis configuration. The rest of the satellites uses tetrahedral configuration (noted as 4 as shown in **Table 1**).

The TU Berlin's attitude control design was chosen to reduce computational burden for filtering out reading noise/jitter in the attitude control sensors. The TU Berlin heritage satellites offer two options for attitude control mode, in addition to regular closed loop, including (1) interactive mode for the satellite with video camera payload, such as DLR-TUBSAT and LAPAN-TUBSAT, and (2) angular momentum management mode for the satellite with line imagers, such as Maroc-TUBSAT and LAPAN-A3. The angular momentum management mode is supported by their structure design, that is, solid aluminum box, which created maximum inertia properties at 1 axis and very little cross-product inertias [30, 31]. Such design has been successfully performed highly stable open-loop angular momentum management operation as published by Utama [31] and Mukhayadi [32].

4.4 Propulsion subsystem

From a selected set of satellite designs shown in **Tables 1–4**, only BILSAT-1 and STSAT-3 have thrusters. The objective for BILSAT-1 thruster is to maintain the satellite orbit separation in the constellation, so that the image coverage could be optimized. In STSAT-3, the plasma thruster is part of in-orbit qualification process for the low power plasma thruster technology developed by KAIST.

4.5 TTC

For Telemetry and Telecommand, the University of Surrey heritage satellite stopped using low frequency (UHF and VHF) after KITSAT-3. Such usage in RASAT is only in emergency situation. Meanwhile, in the TU Berlin heritage, UHF TTC is still used until LAPAN-IPB. The advantage of using low frequency for TTC is on its omni-directional antenna. Therefore, the satellite can always be contacted by its ground station, regardless of its attitude. The cost of the satellite's control ground station is also much lower. However, the risk for frequency noise for its operation is also higher.

4.6 Payload

The payload profiles for both satellite design heritages showed that the platforms are suitable for both Earth observation, science, and low data rate communication missions. All of the selected satellites, except Korean STSAT-1, are Earth observation missions, which are considered important by stakeholder of satellite developer in Korea, Turkey, and Indonesia. KITSAT-3, BILSAT-1, RASAT, and LAPAN-IPB are for land cover that can be applied for estimating crop yield. The payload data showed that combining mission is typical for microsatellite applications. The multiband infrared (MIRIS) payload in STSAT-3 is used for Earth and space observation. LAPAN-ORARI has three kinds of missions, including Earth observation, communication, and ship data collecting platform.

4.7 Mission data downlink

The quantity and quality of the payload in **Tables 1–4** showed that mission data are increasing with time, which increase the required downlink data rate. For the University of Surrey heritage, the data rate started with 3 Mbps in KITSAT-3 and increased to 100 Mbps in RASAT. For the TU Berlin heritage (the digital transmission cases), the data rate started with 256 kbps in Maroc-TUBSAT and increased to 100 Mbps in LAPAN-IPB. In the early missions, the mission data downlink is transmitted in S-band, and as the data rate requirement increases, the downlink has been shifted to X-band.

4.8 Payload computer

Payload computer is typically separated from satellite main computer, which mainly manage the satellite bus. As the payload data rate increased, the payload processing electronics is also evolved, from microcontroller/microprocessor to FPGA based, which is known to be able provide high computing power with less risk from space radiation as compared to high capacity microprocessor.

4.9 Orbit determination

None of the microsatellite has ranging system. Therefore, in early missions, their orbit determination is mainly depending on NORAD's data. The use of GPS for Position-Navigation-and-Timing by the University of Surrey heritage satellites started with BILSAT-1, while for the TU Berlin heritage satellites, it started with LAPAN-ORARI. The accuracy of orbit determination becomes crucial in Earth observation mission, as part of the parameters used in satellite image geometric correction.

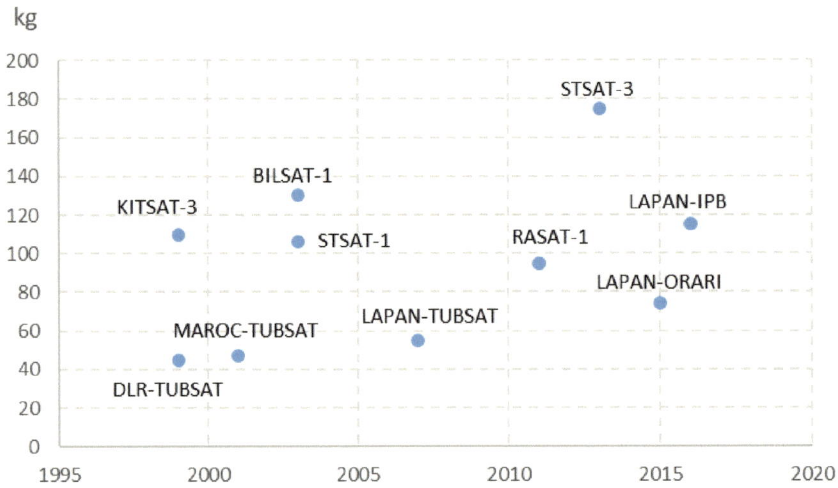

Figure 6.
Microsatellites' weight versus launch year.

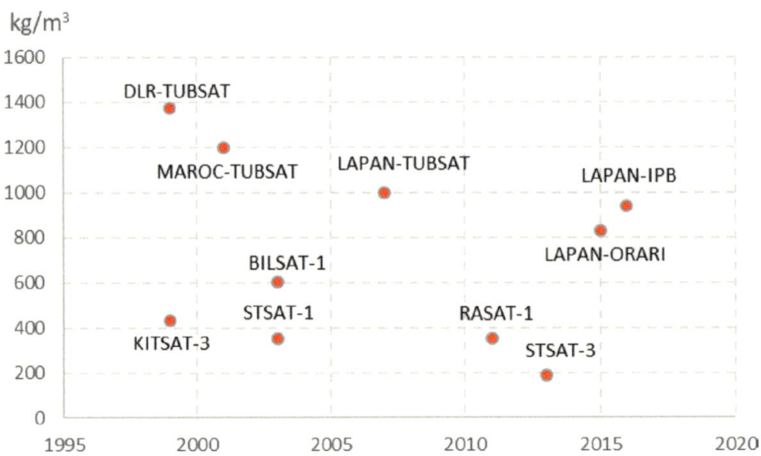

Figure 7.
Microsatellites' density versus launch year.

4.10 System level parameters

Figure 6 shows the weight of each microsatellite sample. It shows that the weight of TU Berlin heritage satellites grows in time. This is due to the increase in mission quantity and complexity, which therefore requires more components in the satellites (bigger batteries, more attitude sensors, larger lens for imager payload, etc.). For the University of Surrey satellites heritage, such pattern is not found. The density of (weight/volume) the satellites is shown in **Figure 7**, indicating that the TU Berlin heritage satellites are more compact than the University of Surrey heritage satellites. For the University of Surrey satellites heritage, the design uses maximum volumetric envelope for maximizing the solar panel area.

5. Conclusions

The chapter has discussed the differences between the University of Surrey design heritage microsatellites and the TU Berlin heritage microsatellites. Five sample satellites from each satellite design heritage are compared, including 15 bus parameters, payload profiles, and satellite weight and volume at launch. From the comparison, it is found that major differences in the satellite bus are in the choice of main computers and their associated link configuration and in the attitude control modes that also affect the design. Another major difference is in the satellites' structure design, which resulted in much higher density in the TU Berlin heritage satellites than the University Surrey heritage satellites. In the early design, there are differences in the choice of satellite's batteries. However, as soon as Li-ion batteries became available, both design heritages used such technology. In answering the increasing needs in payload data handling, both design heritage use FPGA-based payload data handling and high downlink data rate in X-band. GPS is also the technology adopted by both design heritages for orbit determination and imager's ancillary data.

For further studies on the topic, it is suggested that comparison to be done on the power budget of the satellites and on the operation performance parameters of the satellites with similar missions.

Acknowledgements

The author wishes to acknowledge the Center for Aerospace Policy Studies of LAPAN for funding this publication and also wishes to thank the book editor for giving substantial advice in writing this chapter.

Author details

Triharjanto Robertus
National Institute of Aeronautics and Space, Jakarta, Indonesia

*Address all correspondence to: robertus.heru@lapan.go.id

References

[1] Swartout M. Cheaper by the dozen: The avalanche of rideshares in the 21st century. In: 2013 IEEE Aerospace Conference; Big Sky, MT, USA; 2013

[2] Bunchen E. 2014 Nano/Micro-satellite market assessment. In: Proceedings of the 28th AIAA/USU Conference on Small Satellites. Utah; 2014

[3] Kramer HJ. DMC-3. 2016. Available from: https://directory.eoportal.org/web/eoportal/satellite-missions/d/dmc-3

[4] Kramer HJ. SkySat. 2016. Available from: https://directory.eoportal.org/web/eoportal/satellite-missions/s/skysat

[5] Axelspace. GRUS-1: The Next Generation of Earth-Observation Satellites. 2016. Available from: https://www.axelspace.com/en/solution_/grus/

[6] Quigley JT. Japanese Startup Raises $15M to Build Satellites that Are the Size of your Air Conditioner. 2015. Available from: https://www.techinasia.com/japan-micro-satellite-startup-axelspace-series-a

[7] Henry C. How OneWeb Plans to Make Sure Its First Satellites Aren't Its Last. 2019. Available from: https://spacenews.com/how-oneweb-plans-to-make-sure-its-first-satellites-arent-its-last/

[8] Clark S. Telesat Taps Blue Origin to Launch Broadband Satellite Fleet. 2019. Available from: https://spaceflightnow.com/2019/02/04/telesat-taps-blue-origin-to-launch-broadband-satellite-fleet/

[9] Triharjanto R. Strategic environment and implementation of satellite technology acquisition programs. In: ASEAN, Proceeding Seminar Nasional Kajian Kebijakan Penerbangan dan Antariksa; Surabaya, Indonesia; 2018

[10] Sun W, Sweeting MN. In-orbit results from UoSAT-12 Earth observation minisatellite mission. In: Proceedings of the 3rd International Symposium of IAA; Berlin; 2-6 April 2001. pp. 79-82. IAA-B3-0305P

[11] Fouquet M, Sweeting M. UoSAT-12 Minisatellite for high performance earth observation at low cost. Acta Astronautica. 1997;**41**(3):173-182

[12] Lim JT, Nam MR, Ryu KS, Tahk KM, Rhee SH, Kim KH. Exploring space on a small satellite, STSAT-2: A test bed for new technologies. In: 14th Annual AIAA/USU Conference on Small Satellites; 2000

[13] Kim BJ, Lee HW, Choi SD. Three-axis reaction wheel attitude control system for KITSAT-3 microsatellite. Space Technology. 1997;**16**(5/6):191-296

[14] Park JO, Park HY, Lee DH, Lee JH, Rhee SW. STSat-3 (Science & Technology Satellite-3) satellite missions. In: Proceedings of the 60th IAC (International Astronautical Congress); Daejeon, Korea; 12-16 October 2009. IAC-09.B4.2.6

[15] Shin GH, Park HY, Seo JK, Chung TJ, Myung NH. Introduction to micro-satellite STSAT-3 development. In: Proceedings of the 60th IAC (International Astronautical Congress); Daejeon, Korea; 12-16 October 2009. IAC-09.B4.6A.2

[16] Gomes LM, Yuksel G, Lappas V, da Silva Curiel A, Bradford A, Ozkaptan C, et al.. BILSAT: Advancing smallsat capabilities. In: AIAA/USU Conference on Small Satellites; Logan, UT; 11-14 August 2003

[17] Bradford A, Yuksel G, Gomes LM, Ozkaptan C, Sweeting M, Orlu U. BILSAT-1: A low-cost, agile, earth observation microsatellite for

Turkey. In: Proceedings of 53rd IAC; Houston, TX; 10-19 October 2002. IAC-02-IAA.11.3.02

[18] Leloglu UM, Yuksel G, Sweeting MN. BILSAT-1: A case study for the Surrey Satellite Technology Ltd Know-How Transfer and Training Program. In: Proceedings of 53rd IAC; Houston, TX; 10-19 October 2002. IAC-02-P.1.01

[19] Yuksel G, Belce O, Urhan H, Gomez L, Bradford A, Bean N, et al. BILSAT-1: First year in orbit-operations and lessons learned. In: Proceedings of AIAA/USU Conference on Small Satellites; Logan, UT; 9-12 August 2004. SSC04-IX-3

[20] Kim EE, Choi YW, Choi W, Kim HG, Park SJ, Yun JH, et al. Integration and testing of optical imaging system for RASAT. In: Proceedings of Recent Advances in Space Technologies (RAST); Istanbul, Turkey; 2007

[21] Ontaç S, Dağ S, Gökler MI. Structural finite element analysis of stiffened and honeycomb panels of the RASAT satellite. In: Proceedings of the 3rd International Conference on Recent Advances in Space Technologies (RAST 2007); Istanbul, Turkey; 14-16 June 2007

[22] Yüksel G, Okan A, Leloğlu UM. First LEO satellite built in Turkey: RASAT. In: Proceedings of the 3rd International Conference on Recent Advances in Space Technologies (RAST); Istanbul, Turkey; 14-16 June 2007

[23] İsmailoğlu N, Vargün B, Sever R, Okcan B, Karakuş K, Kapucu K, et al. GEZGN-2: Image processing subsystem of RASAT. In: Proceeding 5th International Conference on Recent Advances in Space Technologies (RAST); 2011

[24] Schulz S, Renner U. DLR-TUBSAT, A microsatellite for interactive earth observation. In: Proceeding 5th International Symposium Small Satellite Systems and Services; La Baule, France; 2000

[25] Renner U, Bleif J, Roemer S. The TUBSAT attitude control system: flight experience with DLR-TUBSAT and MAROC-TUBSAT. In: 4th IAA Symposium; Berlin, Germany; 7-11 April 2003. IAA-B4-1201

[26] Roemer S, Renner U. Flight experience with the microsatellite MAROC-TUBSAT. In: 54th IAF Congress; Bremen, Germany; September 29–October 3, 2003; IAC-03-IAA.11.2.07

[27] Triharjanto RH, Hasbi W, Widipaminto A, Mukhayadi M, Renner U. LAPAN-TUBSAT: Micro-satellite platform for surveillance & remote sensing. In: Proceedings of the 4S Symposium: Small Satellites, Systems and Services; La Rochelle, France; 2004

[28] Saifudin MA, Mukhayadi M. LAPAN-A2 attitude control strategy for equatorial surveillance mission. In: Proceedings of the 9th IAA Symposium on Small Satellites for Earth Observation; Berlin, Germany; 8-12 April 2013. Paper: IAA-B9-1202

[29] Triharjanto RH, Hakim PR. Review of satellite technology development in Indonesian Space Agency based on its technical publications in 2012-2016. In: Proceedings of the 68th International Astronautical Congress; Australia; 2017

[30] Triharjanto RH. The comparison of LAPAN-A1 and LAPAN-A2 system design: Weight ratios and moment inertias analysis. In: Proceedings 2nd International Seminar of Aerospace Science and Technology; Jakarta, Indonesia; 2014

[31] Utama S, Saifudin MA, Mukhayadi M. Momentum biased performance of LAPAN-A3 satellite

for multispectral pushbroom imager operation. IOP Conference Series Earth and Environmental Science. 2018;**149**(1):012062

[32] Mukhayadi M, Madina R, Renner U. Attitude control of bias momentum micro satellite using magnetic and gravity gradient torque. In: Proceedings of the 2014 IEEE International Conference on Aerospace Electronics and Remote Sensing Technology; Yogyakarta, Indonesia; 2014. pp. 132-136

Chapter 7

Design of Intelligent and Open Avionics System Onboard

Changqing Wu, Xiaodong Han and Yakun Wang

Abstract

The continuous development of space missions has put forward requirements for high performance, high reliability, intelligence, effective integration, miniaturization, and quick turn around productization of the electronic system of satellites. The complexity of satellites has continued to increase, and the focus of satellite competition has shifted from the launch of success shifts to communication capacity, performance indicators, degree of flexibility, and continuous service capabilities. So, the importance of onboard avionics system is becoming increasingly prominent. In the future, the advanced avionics system integrates most of the platform's electronic equipment. The design level of the system largely determines the performance of the satellite platform. This chapter focuses on the application requirements of the new generation of intelligent avionics system for future communication satellites and adopts an "open" architecture of "centralized management, distributed measurement and drive, and software and hardware 'modular' design" to build a universal, standardized, and scalable intelligent avionics system.

Keywords: satellite, avionics system, intelligent, open architecture, modular design, centralized management, reliability

1. Introduction

With the continuous advancement of electronics and computer technology, the functions and performance of spacecraft avionics system have also continuously improved, covering functions such as spacecraft remote measurement and remote management, energy management, thermal management, health management, payload information processing, and mission task management. Avionics system plays a core role in the realization of information sharing and comprehensive utilization, function integration, resource reorganization and optimization, and information processing and transmission [1]. It is the foundation for spacecraft to implement autonomous management and control and is also a bridge for communication management from a spacecraft to other spacecrafts and ground station [2].

The traditional spacecraft electronic system uses a layered centralized management control mode similar to a pyramid. It not only needs a large amount of data interaction between the management unit and the interface unit but also requires the management unit to process a large amount of underlying data, which makes the management unit overwhelmed. It severely limits the processing and support of high-level tasks by electronic systems. Moreover, the management unit is at the top of the "pyramid" of centralized management, which requires high reliability. Once

 IntechOpen

a failure occurs, the entire electronic system will fail. Thus, the centralized management method is no longer suitable for the needs of spacecraft development.

The satellite intelligent avionics system is an information processing and transmission system that uses computer network technology to interconnect satellite-borne electronic equipment on the satellite to achieve internal information sharing and comprehensive utilization, function integration, and resource reorganization and optimization. Utilizing onboard computers to complete satellite data management, control management, communication management, time management, energy management, and job management functions through unified scheduling of satellite missions. Its essence is the generation, identification, processing, analysis, transmission, and distribution of information process. The integrated satellite electronic system integrates the functions of the satellite platform electronic equipment, and its design level directly determines the performance of the satellite platform [3–5].

At present, satellite sub-systems mostly adopt independent design schemes, which decentralize satellite attitude control, propulsion control, thermal control, satellite-ground link communication, and power control functions. The onboard computer is responsible for tasks such as remote control, telemetry, program-controlled operation, thermal control, and time management. The attitude and orbit control computer are responsible for attitude and orbit (including propulsion control) control. Each sub-system such as power supply, thermal control, and digital transmission is equipped with corresponding lower-level computers responsible for telemetry acquisition and remote control of the respective sub-system. However, the satellite system designed using this approach is usually resulting in heavy weight, high power consumption, large volume (aka high size, weight, and power (SWAP)), complex interface relationships, weak system reconfiguration capabilities, and low functional density. In order to overcome the abovementioned shortcomings and make the satellite avionics system better meet the SWAP and flexible system configuration requirements of future missions, it is necessary to improve its design technology, that is, from the current independent design of each sub-system to the open and modular design of the entire satellite. Based on the principle of unified application, deployment and operation of hardware resources, and the full use of the various functions of the software, the information sharing of the entire satellite, simple system configuration, and overall performance optimization are realized.

This chapter focuses on the application requirements of the new generation of intelligent avionics system for future communication satellites, and adopts an open architecture of "centralized management, distributed measurement and drive, and software and hardware modular design." The universal, standardized, and scalable intelligent avionics system is built based on the basic modular elements of open hardware modules, open software components, and industry standardized internal and external busses.

2. System structure

This section introduces the intelligent open system architecture, including Sections 2.1, 2.2, 2.3, and 2.4. Section 2.1 introduces the overall architecture design; the system adopts the distributed design mode and completes the intelligent management of onboard tasks through the menu hardware architecture and open interface protocol. Section 2.2 discusses the hardware architecture of high-performance computing and introduces the onboard high-performance computing and the corresponding storage capacity from the main functions, processing, storage, and radiation resistance. Section 2.3 describes the dynamic state reconfigurable task scheduling that improves the fault tolerance ability of the satellite network in view of the typical scenarios of the satellite integrated electronic system in the operation process. Section 2.4 discusses the design of software partition protection mechanism related to the next-generation avionics system and analyzes the requirements, design, and functions of partition protection, aiming to improve the robustness of the software system.

2.1 Overall architecture design

The architecture of the newly proposed next generation of intelligent communication satellite avionics system is shown in **Figure 1**. The avionics system architecture (ASA) is designed as a data bus (DB)-based real-time distributed computer system. ASA consists of one Satellite Management Unit (SMU), one Platform Integrated Services Unit (PFISU), one Payload Integrated Services Unit (PLISU), and a set of DB and auxiliary software. The SMU is the core of the avionics system. ASA controls the PFISU and PLISU by DB and connects with Telemetry and Telecommand Unit (TTU) to receive commands and send the telemetry data. PFISU and PLISU are the execution parts of the avionics system. PFISU and PLISU are used to command driver, signal sample, power distribution, heater control, pyrotechnic management, and interface management. To improve the reliability of avionics system, the SMU, PFISU, and PLISU will have built in redundancies. This avionics system supports the functions of satellite on-orbit dynamic registration, spatial data interaction, and routing and can solve the problem of user-oriented and task-oriented opening of satellite system.

2.2 High-performance computing hardware architecture

As the amount of data generated by satellite electronic equipment continues to increase, a large amount of data processing requirements place higher requirements on satellite information processing capabilities. The avionics system is the

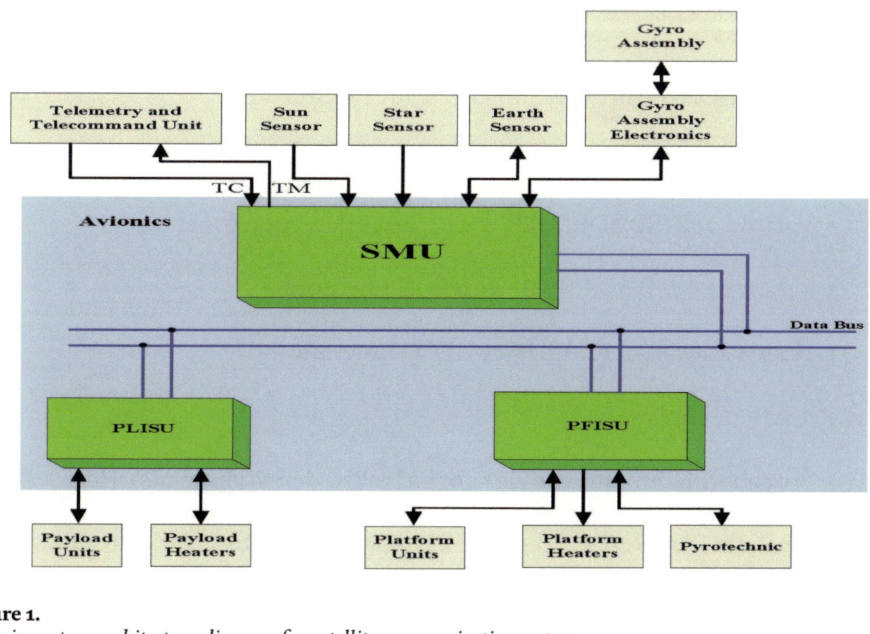

Figure 1.
Avionics system architecture diagram for satellite communication system.

information core of the whole satellite, especially for the requirements of intelligent satellite systems. Research on the realization of high-performance computing of avionics systems is an inevitable requirement [6, 7]. In order to improve the computing capacity of the avionics system, a high-performance onboard processor is utilized. The following introduces onboard high-performance computer from four aspects: main functions, processing, storage, and anti-irradiation.

i. Main functions:

- Uses redundant onboard computer supporting on-orbit reconstruction and reconfiguration for highly reliable avionics system.

- Supports Consultative Committee for Space Data Systems (CCSDS) telemetry and telecommand with optional radio-frequency channels. This feature allows the proposed intelligent avionics system design to be open interface using widely acceptable industry standards.

- Uses interface with external unit. Provide a brief description why is this important function.

- Provides secondary power distribution and discrete instructions to external units. Provide a brief description why is this important function.

ii. High-performance processors:

- 215 Dhrystone Million Instructions executed Per Second (DMIPS) and floating-point arithmetic unit

- L1 instruction cache and L1 data cache with Error Correcting Code (ECC) function

- Internal Random Access Memory (RAM), FLASH, and Electrically Erasable Programmable Read-Only Memory (EEPROM), with ECC function

- Contains basic software: BIOS and startup software

iii. High-performance memory:

- Volatile: 192 MB SDRM CPU, with error detection function

- Volatile: 64 MB SDRM IO, with ECC function

- Nonvolatile: 4GB FLASH, with ECC function

iv. Radiation resistance:

- Spaceborne components will not be locked due to space radiation.

- Dual-core processor lockstep technology is used for error detection.

- All memories have ECC function (RS code or EDAC).

Among them, the "lockstep" technology is a fault-tolerant computing technology. This technology uses the same, redundant hardware components and processes the same instructions at the same time. The core idea is to keep multiple central processing units (CPUs) and memories executing the same instructions accurately and synchronously by running synchronous comparisons in operation to improve the fault-tolerant computing capability of the avionics system.

2.3 Dynamically reconfigurable task scheduling

The two typical scenarios usually encountered by satellite avionics systems during operation are (a) a node fails or requires functional reorganization so that some tasks on this node need to be migrated to other nodes through the network and (b) the resource occupancy rate of a node is too high so that some tasks on this node will be migrated to other relatively idle nodes for execution. The avionics system is designed with networked real-time multitasking distributed system software, which can also implement dynamic reconfiguration of functions and task scheduling. The embedded system software running on each node in the network supports not only the local real-time multitasking scheduling but also the network operation capability. The avionics system supports function modification and function migration between nodes, which realizes the transformation to software-defined satellite functions, reduces the differences in hardware products, improves the fault tolerance of the intra-satellite network, and also meets the growing needs of intra-satellite networking [8].

The avionics system networked real-time multitasking distributed system software is shown in **Figure 2** and has the following characteristics:

- Application tasks are directly oriented to users. In order to complete a top-level function in a specific domain, the tasks are decomposed into functions of appropriate granularity. The software functions that multiple tasks will use are called domain public services. It is called public services in multiple fields, has a clear interface definition, can complete certain functions relatively independently, and adds service registration, management, control, and governance to provide strong support for space application tasks. The user's service

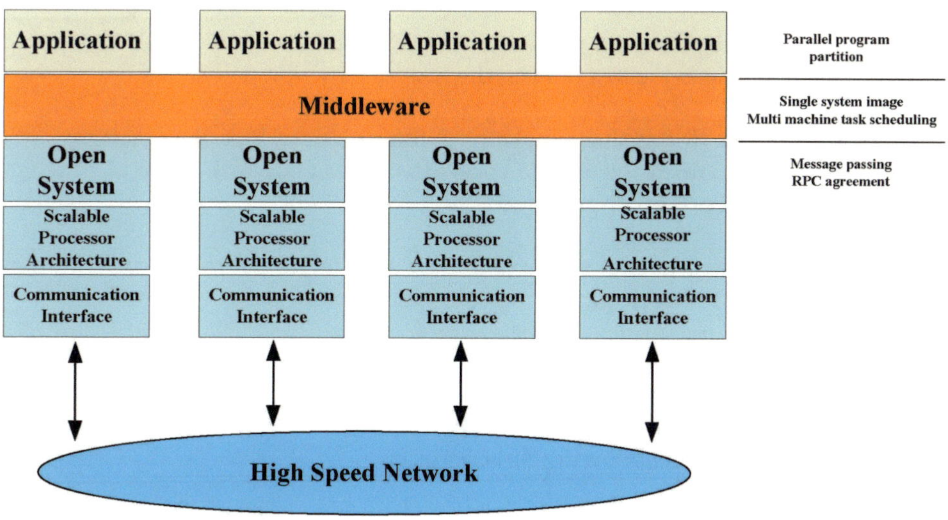

Figure 2.
Networked real-time multitasking distributed system software.

composition capability, that is, when the business needs of the day change and the service call is adjusted, can support the user to quickly combine services and form a new business process [9].

- The middleware layer serves as a transition layer between the above and the next [10]. By combining the characteristics of the tasks of each layer, the corresponding theme is designed, and the publish-subscribe technology is used to provide the entire application layer with access to various resources in the basic resource layer. The software layer implements operating system and communication protocol level shielding through the packaging of thread tasks, synchronization resources, memory access, IO operations, Ethernet, shared memory, and fiber-optic communications, providing access to the hardware device layer. By virtualizing the calculation, storage, and network resources of the basic resources, the basic resources as a service are realized, and the availability and scalability of the hardware resources are guaranteed. At the same time, the software is dynamically deployed for the hardware of the basic resources, including the automation of basic software and application software installation settings, maintenance and upgrades, etc., and provides the system with general basic services such as system reconstruction, software fault tolerance, data management, subscription release, etc.

2.4 Software partition protection mechanism design

i. Partition protection requirements

Avionics systems can implement multiple functions to share resources. The functional entities (which can be software modules, hardware modules) that share resources are called partitions. The partitions of the original avionics system shared resources, but sharing would bring potential problems described below:

- Multi-partition shared memory and IO: if one partition accidentally or maliciously rewrites the memory and IO of other partitions, it will cause the rewritten partition to fail.

- Multi-partition shared processor time: if a partition maliciously takes up processor time due to a failure, the related partition will crash.

- Multiple partitions share the same communication link: if a partition occupies channels too much, it will affect the bandwidth and real-time performance of other partitions.

Therefore, the design of system software architecture of the avionics system should meet the reliable partition protection to avoid the above problems.

ii. Partition protection design

The partition protection design of onboard system software in avionics systems includes the following three aspects:

- Space protection. For processors with Memory Management Unit (MMU) support, such as X86 processors, it is stipulated that the partition itself cannot directly access physical memory, and only virtual memory can be accessed through the MMU memory mapping table configured for each partition by system software. For processors without MMU support, taking the Scalable Processor Architecture (SPARC) processor as an example, during partition initialization, the system software protects the SRAM by setting the privileged register to achieve partition space isolation.

- Inter-partition communication. Inter-division communication is another important content of partition space protection. Inappropriate inter-partition communication mechanisms can cause mutual interaction between partitions. ARINC-653 specification provides a communication mechanism that does not affect each other. Inter-partition communication within a single processor includes a memory buffer mechanism and a blackboard mechanism. The buffer mechanism mainly provides data communication between partitions, and the blackboard mechanism mainly provides sampling services between partitions. Taking the buffer mechanism as an example, the protection mechanism for inter-partition communication is the following: (1) The core system software allocates an inter-partition communication buffer (size, permission, and connection relationship between buffers) for each partition according to the blueprint information. Only the own buffer can be accessed. (2) The core system software manages the resources for communication between partitions in the privileged state and copies the information in the source buffer to the destination buffer according to the blueprint information.

- Partition time protection. Its strategies include the following: (1) The basic unit of scheduling is partitioning, and partition scheduling has no priority. The Main Time Frame (MTF) is used to statically define the scheduling order of each partition and the proportion window size. MTF is one of the blueprint contents. The core system software is configured according to the MTF, and multiple partitions are scheduled in a cyclic manner. (2) The resources (such as timers, stacks, and memory) and blueprints required for partition scheduling are in a privileged state, and the user partition cannot overwrite the partition scheduling resources.

iii. Advantages of partition protection

- Blueprint only registers the interfaces on the modules and then mounts them on the app as a whole. The purpose of blueprint itself is to organize the parallel coexistence of multiple modules and avoid registering modules directly on the app. In fact, it is more convenient for development and code maintenance, because ultimately all interfaces on views are still directly mounted on the app, which corresponds to the entire application; there is no obvious difference [11].

- Blueprint is not a pluggable application, because it is not a real application, but a set of operations that can be registered in the application and can be registered multiple times.

- At the same time, we cannot use multiple objects to manage and register, because this will cause each object to have its own configuration, which is not easy to manage.

- With blueprint, the application will be managed in the flask layer, share the configuration, and change the application object on demand through registration. The disadvantage of blueprint is that once an application is created, it can only be unregistered by destroying the entire application object.

3. Menu system composition

Based on the menu design idea, the avionics system can realize the sharing of hardware modules and resources, task migration, and system reconstruction, enhance the tolerance and processing ability of the avionics system for faults, significantly improve the development efficiency and productization degree of the integrated electronic system, and make the integrated electronic system highly reliable. The intelligent spacecraft provides the necessary technical support, mainly including the contents that will be presented in the subsequent sections [12].

3.1 Satellite management unit

The Satellite Management Unit is an improved satellite research equipment. The design fully draws on the advantages of the previous satellite platform and has been optimized and expanded. The main completed functions are as follows:

i. Telecommand function: The SMU receives the telecommand instructions from the TTU and completes the distribution of the instructions.

ii. Telemetry function: The SMU collects its own telemetry and receive the indirect telemetry parameters collected by the bus terminal equipment through the 1553B bus. The SMU complete the framing processing according to the CCSDS standard and transmit it to the TTU through the serial port.

iii. Satellite autonomous management function: Using application software running on the SMU, functions such as energy management, thermal management, bus management, payload management, and pyrotechnics management are realized.

iv. On-orbit maintenance function: For the temporary adjustment of control parameters during the execution of on-orbit tasks, the SMU is able to modify the control parameters of the software. It also supports onboard software maintenance function of onboard software, which can realize the update and recovery of software modules.

v. Important data saving function: The SMU can periodically save and maintain important data of satellite sub-systems. When the internal configuration of the relevant sub-system changes or the corresponding module restarts, the SMU can send the important data stored internally to the corresponding module. Besides, when the SMU is reset or switched off, the SMU can restore the current working state through the important data.

vi. Fault Detection Isolation and Recovery (FDIR) function: The SMU provides the operating platform for satellite FDIR [13]. When the SMU is healthy, the SMU monitor key information such as the entire satellite's energy and thermal control. The SMU detects various failure conditions in real time and performs troubleshooting through the direct remote command interface or through the 1553B bus. The FDIR of the control sub-system is completed by the attitude control computer. Note that the most advanced 1553B bus can handle data rate up to 10 Mbps. For data rate higher than 10 Mbps, industry trend is moving toward SpaceWire data bus that can handle data rate up to 400 Mbps.

The SMU adopts the dual-machine cold backup working mode. In the case of the autonomous switching enabled, when the on-duty machine fails, the failure-tolerant module automatically completes the switch of SMU and power off the faulty machine. In the case of the autonomous switching not enabled, when the on-duty machine fails, the power failure and machine switch is conducted by ground station.

3.2 Integrated service unit

The Integrated Service Unit (ISU) adopts modular hardware design [14, 15]. Each functional module is connected to the "bus interface management module" through an internal bus, which uses widely acceptable data bus such as 1553B bus. The ISU is mainly composed of a bus interface management module, a matrix instruction and matrix telemetry module, and an analog quantity acquisition and a discrete instruction output module.

The menu module is mainly composed of a bus interface management module, a matrix acquisition and command module, and an information acquisition and command module [16]. The module menu is shown in **Table 1** below.

The functions of each module are also modularized. The capability of each module is shown in **Table 2**.

After the module design is completed, the number of modules is determined according to the task requirements. Then, complete the assembly according to the standard interface. The number of modules menu is shown in **Table 3**.

3.3 Data bus network

The data bus network is the information transmission hub of the avionics system. Through the data bus network, distributed data acquisition, and instruction output, centralized operation and control are implemented, thereby improving the efficiency of system processing. The avionics system first-level bus is 1553B bus [17]. Note that for data rate higher than 10 Mbps, SpaceWire data bus is recommended.

Data exchange between SMU and ISU and other equipment realized through 1553B bus. The master–slave response mode of the 1553B bus is used to transmit platform command data and telemetry acquisition data. In the data exchange process of the first-level bus, the SMU always acts as the controller of the 1553B bus

Functions	Module menu
Telemetry/telecommand	Acquisition and Command-A module (AA)
	Acquisition and Command -B module (AB)
	Matrix acquisition and command module (AC)
Temperature acquisition	High-voltage heater control module (HH)
	Low-voltage heater control module (LH)
	Heater and distribution module (HD)
Pyrotechnic management	Pyrotechnic management module (CA)
Bus data transmission	1553B bus

Table 1.
Function module list.

Module	Module capabilities
AA	Analog/temperature measurement acquisition channel
	Small current discrete command output
	High current command drive circuit
AB	Analog/temperature measurement acquisition channel
	Small current discrete instructions
	Bi-level quantity acquisition
AC	Matrix instructions
	Switch status acquisition
HH	High-pressure heater power distribution
	Temperature measurement collection
LH	Low-voltage heater power distribution
	Temperature measurement
HD	High-voltage instrumentation and power distribution
	High-pressure heater power distribution
	Temperature measurement
CA	Pyrotechnic management

Table 2.
Modularized functions.

Demand	Quantification	AA	AB	AC	CA
Pyrotechnics management	80				2
High current instruction	30	1			
Low current instruction	100		1		
Analog acquisition	220	2	2		
Matrix acquisition	550			2	
Total (take the maximum)		2	2	2	2

Table 3.
Menu-style design.

and initiates the communication. ISU and other equipment, as the RT end, receive instructions and send collected telemetry data to the SMU.

4. Failure detection isolation and recovery (FDIR) design

In order to achieve autonomous and healthy operation of the satellite, the intelligent satellite system uses the FDIR software to monitor the status of the satellite in real time and diagnose and predict its working status and performance trends [18]. When a failure occurs, the FDIR software can locate the failure in time and determine which components are not working normally or the performance is degraded.

4.1 FDIR design goals and principles

Design goals:

i. Satellites can survive if any failure occurs.

ii. When a failure occurs, try to extend the mission time of the satellite and reduce the loss of mission interruption.

iii. The life of the satellite should be guaranteed: optimize fuel consumption and minimize system configuration and component losses.

The above three principles apply to the launch phase, the transfer orbit phase, and the on-orbit phase.

FDIR is an important component of the onboard software, which can perform on-orbit processing of failures, thereby reducing the impact of failures. However, not all on-orbit failures can be detected and processed. FDIR design should follow the following principles:

i. FDIR processing follows the single failure principle, that is, only one failure is processed at a time.

ii. Failures are divided into 0 to 4 levels according to their impact on satellites.

iii. The higher the failure level, the higher the processing priority. During a failure processing, if a higher-level failure occurs, the higher-level failure is processed first.

iv. The failures of the same level are processed in the order of occurrence.

v. All FDIR processing requires failure recovery instructions and failure processing records.

4.2 Failure levels

According to the impact of the failure on the satellite operation, the failure is categorized as follows:

i. System-level failure: failures that damage the functions and performance of the satellite system.

ii. Sub-system-level failure: the functions of the sub-system cannot be or are partially completed, or the main performance indicators and parameter values of the sub-system exceed the range required by the sub-system design. But it does not affect the main functions and performance of the system.

iii. Equipment-level failure: equipment functions cannot complete the main performance indicators, or parameter values exceed the range of equipment design requirements. But it does not affect the main functions and performance of the system.

iv. Module-level failure: a failure in which the module function cannot be completed, or the main performance indicators and parameter values exceed the range required by the component design. But it does not affect the main function and performance of the equipment.

According to the possible impact on components, functions, and systems, FDIR is designed for five failure levels from levels 0 to 4, according to the different sub-systems that each failure belongs to, including measurement and control FDIR, avionics FDIR, and power supply and distribution FDIR as illustrated in **Figure 3**. The larger the number, the higher the fault level, and vice versa.

- Level 0 failure: a level 0 failure refers to a failure that occurs inside an equipment and can be recovered autonomously by the hot backup method inside the equipment without affecting other components of the system.

- Level 1 failure: a level 1 failure refers to the failure of a single equipment or module of each sub-system. After a level 1 failure occurs, the system will perform autonomous failure isolation and recovery according to the FDIR policy. If the failure isolation and recovery is successful, it has no impact on system tasks. The detection, isolation, and recovery of failures are implemented by application software.

- Level 2 failure: a level 2 failure refers to the functional level abnormality of the satellite sub-system. Under such failures, the system performance cannot meet the design requirements. For level 2 failures, the recovery strategies need to be implemented and related components need to be enabled or restarted. Level 2 failures can cause system performance degradation or temporary interruption of system tasks. Its failure detection, isolation, and recovery are performed by application software.

- Level 3 failure: a level 3 failure refers to the failure of the CPU hardware, which is detected by the hardware. After the failure occurs, it is switched to the backup CPU according to the failure handling strategy.

- Level 4 failure: a level 4 failure refers to the failure of the satellite to maintain the pointing to the ground in the on-orbit phase and requires sun capture processing.

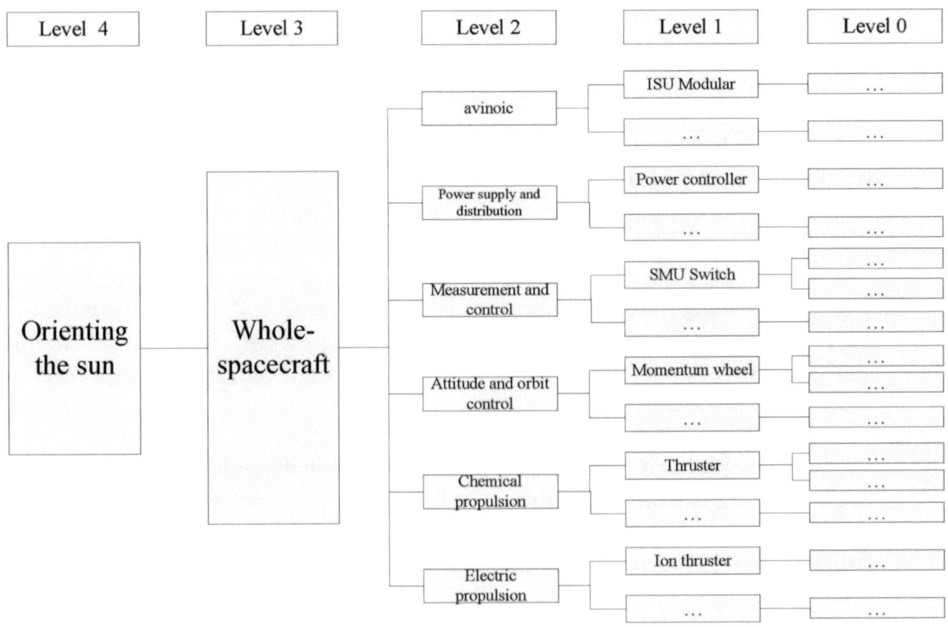

Figure 3.
Schematic diagram of satellites in orbit during their lifetime.

4.3 FDIR scheme

The software autonomously isolates the failure and rebuilds the system at the appropriate time according to the following FDIR scheme:

- High-level failure detection such as level 4 failure has priority over low-level failures such as level 3. When two or more failures are detected at the same time, the recovery sequence of high-level failures is performed preferentially. Once a high-level failure occurs, all detection of the same-level and low-level failures are suspended before the recovery sequence is completed.

- Only one failure recovery sequence is performed on the satellite at the same time, that is, all FDIR failure recovery is shielded during the execution of any failure recovery strategy. The sufficiency and necessity of the failure recovery sequence should be effectively verified by ground testing to minimize the interpretation during the sequence execution.

- After the execution of the failure recovery sequence is completed, the failure detection of the flight can be continued, but the failure recovery enable status should be set to disable. At the same time, the detection of other similar and low-level failures should be enabled. After confirming the working status of the products on the ground, reset the backup status and enable the FDIR recovery function.

- FDIR only detects the status of the on-duty module. When the FDIR enable flag is "disabled," the status of the module is not detected. When the FDIR enable flag is "enabled," the status of the on-duty module is detected. If the failure detection condition is met on the on-duty module and the FDIR recovery enable flag is "disabled," the health status of the module is set to unhealthy. If the failure detection condition is met on the on-duty module and the FDIR recovery enable flag is "enabled," it is determined whether the status of the on-duty module is the same as the backup module. If they are the same, do not perform the recovery operation and set the on-duty module as unhealthy. If they are different, perform the recovery operation and set the backup module to the on-duty status.

- For autonomous maintenance on the satellite, the "health status" of each module can only be changed from "healthy" to "unhealthy."

- For dual-machine hot standby equipment or modules, only the health status of non-duty module is detected, and no recovery is performed.

- Use its own fault-tolerant RAM and lower computer to save important data in time for state recovery after failure.

4.4 FDIR processing requirements for satellites in orbit

The FDIR requirements for each phase of the satellite are as follows:

i. Launch phase: allows failure detection and recovery of level 0 and level 1 failures.

ii. Transfer orbit phase: allows failure detection and recovery of level 0 to level 3 failures.

iii. On-orbit phase: allows failure detection and recovery of level 0 to level 4 failures.

4.5 FDIR processing

The processing flow of FDIR mainly includes four parts:

i. Judgment of processing conditions: First, determine the scope of failure detection according to the requirements of the satellite in orbit and ground control. Then, according to the validity of the telemetry data and the situation of the modules on duty, determine the FDIR project that can be used for failure detection.

ii. Fault detection: Determine whether a failure occurs based on the recognition characteristics.

iii. Comprehensive information processing: After a failure occurs, it is determined that whether the current situation of the satellite meets the recovery conditions. At the same time, in the case of multiple failures, priority judgment is required. Finally, determine the failures that can be recovered and the order of recovery.

iv. Failure recovery: According to the engineering and testing experience, perform corresponding recovery operations.

5. Impacts on next-generation avionics system

The intelligent avionics system adopts a system engineering method using modular and open design to uniformly design the information processing, control and management processes, hardware, and software, which is to realize the optimization of information and resource sharing. Based on the onboard computer and high-speed bus such as SpaceWire data bus, a set of information fusion systems and mechanisms is established. The system is a menu-style, modular, and extensible open service platform, which achieves a high degree of integration of various onboard software and hardware resources and can meet the requirements for different tasks.

The intelligent avionics system adopts the design concept of a modular menu system architecture to meet the needs of real-time, reconfigurable, autonomous planning, and intelligence of the system. With the SMU as the 1553B and SpaceWire bus controller for data rate less than 10 Mbps/for data rate more than 10 Mbps, respectively, and the ISU as the remote terminal, a distributed, master-slave, and menu-based satellite networks are constructed.

Satellites are designed with a network layout, which can design different menu network nodes on the bus network. After the payload capacity is strengthened, the network node can increase the corresponding payload processing unit. The SMU is used as the main processing computer to perform the main control of satellite services to form a master-slave network structure. The high-performance onboard processor enables the intelligent avionics systems with high-performance computing capabilities, which not only meets the data processing requirements but also lays the foundation for satellite intelligence. The intelligent avionics system adopts partition protection measures. Through the design of space protection, time protection, and partition communication, it provides reliable functional entities (such as software modules or hardware modules) that share resources. Partition protection avoids the impact of other partitions under abnormal conditions such as single partition failure or malicious access. At the same time, the intelligent avionics system is equipped with real-time multitask distributed system software, which can realize the dynamic reconstruction of functions and tasks. For example, when a node fails or needs a functional reorganization, some tasks on that node will be migrated to other nodes. Or, the resource occupation rate of a node is too high, and some tasks on this node will be migrated to other relatively idle nodes for execution. This design can improve the failure tolerance of the intra-satellite network and achieve efficient resource allocation and scheduling. All information is collected into the SMU for comprehensive analysis and processing through the 1553B/SpaceWire bus network. For example, in the process of autonomous energy management, it is found that the battery discharge depth reaches 80%. If the control sub-system is still in the mode of pointing to the ground, it will seriously affect the safety of the satellite. At this time, the instructions should be sent in time to orient the satellite to the sun to ensure the safety of the satellite. The SMU can be fully applied to the satellite's autonomous information fusion processing, ensuring that the satellite can still guarantee normal communication services in the event of a major failure, and energy security in emergency situations.

6. Conclusion

The intelligent avionics system design is the key technology for future advanced satellites. The system design has adopted modular and open system architecture approach using an efficient computing hardware system to maintain multiple central processing units and memory executing instructions accurately and synchronously with high cost-performance and cost-efficiency ratio. This approach improves the failure tolerance of the next-generation avionics systems. The function modification capabilities and function migration between modules realize the transition to software-defined satellite. Furthermore, this approach also reduces the differences in hardware products and improves the failure tolerance of the satellite's internal network, which meets the ever-increasing networking requirements of the satellite.

Acknowledgements

This work was supported in part by the National Natural Science Foundation of China (No. 61972398).

Author details

Changqing Wu, Xiaodong Han* and Yakun Wang
Institute of Telecommunication Satellite, China Academy of Space Technology, Beijing, China

*Address all correspondence to: willingdong@163.com

References

[1] Sherwood R, Chien S, Tran D, et al. The EO-1 autonomous sciencecraft. In: Small Satellite Conference. Logan, UT: AIAA/USU; 2007

[2] Rabideau G, Tran D, Chien S, et al. Mission operations of earth observing-1 with onboard autonomy. In: IEEE International Conference on Space Mission Challenges for Information Technology. Pasadena, CA: IEEE; 2006

[3] Sherwood R, Chien S, Tran D, et al. Enhancing science and automating operations using onboard autonomy. In: International Conference on Space Operations (SpaceOps 2006). Rome, Italy: AIAA; 2006

[4] Sherwood R, Chien S, Tran D. Autonomous science agents and sensor webs: EO-1 and beyond. In: IEEE Aerospace Conference (IAC 2006). Big Sky, MT: IEEE; 2006

[5] Chien S, Sherwood R, Tran D, et al. Using autonomy flight software to improve science return on earth observing one. Journal of Aerospace Computing, Information, and Communication. 2005;**2**(4):196-216

[6] Chien S, Sherwood R, Tran D, et al. Lessons learned from autonomous sciencecraft experiment. In: Autonomous Agents and Multi-Agent Systems Conference (AAMAS 2005). Utrecht, Netherlands: AAMAS; 2005

[7] Rabideau G, Chien S, Sherwood R, et al. Mission operations with autonomy: A preliminary report for earth observing-1. In: International Workshop on Planning and Scheduling for Space (IWPSS 2004). Darmstadt, Germany: IWPSS; 2004

[8] Marie J. Spacebus 4000 avionics: Key features and first flight return. In: 24th AIAA International Communications Satellite System Conference (ICSSC). 2006

[9] Heymans P, Boucher Q, Classen A. A code tagging approach to software product line development. An application to satellite communication libraries. International Journal on Software Tools for Technology Transfer. 2012;**14**(5):553-566

[10] Boreli R, Ge Y, Iyer T, et al. Intelligent middleware for high speed maritime mesh networks with satellite communications. In: 9th International Conference on Intelligent Transport Systems Telecommunications (ITST). Harbin, China: IEEE; 2009

[11] Zhu H. Design of on-board software architecture based on design pattern. Computer Application. 2008;**25**(12):180-181

[12] Yu Z, Yang-Ming Z, Zheng-Liang H, et al. Fault-tolerant design of memory module for pica-satellite on-board computer. Journal of Astronautics. 2008;**29**(6):2057-2061

[13] Guiotto A, Martelli A, Paccagnini C. SMART-FDIR: Use of artificial intelligence in the implementation of a satellite FDIR. In: DASIA 2003. Prague, Czech Republic: DASIA; 2003

[14] Herpel HJ, Schuettauf A, Willich G. Open modular computing platforms in space — Learning from other industrial domains. In: IEEE Aerospace Conference. MT, USA: IEEE; 2016

[15] Mark H, Helene DM, Alexandre C. Requirements baseline for integrated modular avionics for space separation kernel qualification. In: Dasia-Data Systems in Aerospace. Barcelona, Spain: DASIA; 2015

[16] Panpan Z, Tingyuan G, Jianjun G, et al. Plug-and-play on-board computer system design based on BM3803 processor. Journal of Aerospace Engineering. 2013;**22**(06):92-96

[17] Kan W, Jingfei J, Mianjiang H, et al. Intelligent fault-tolerant 1553B bus system based on adaptive learning. In: IEEE International Conference on Communication Software and Networks. Xi an, China: IEEE; 2011. pp. 378-381

[18] SalarKaleji F, Dayyani A. A survey on fault detection, isolation and recovery (FDIR) module in satellite onboard software. In: International Conference on Recent Advances in Space Technologies (RAST). Istanbul, Turkey: IEEE; 2013

Section 4

Satellite Systems Modeling Simulation and Analysis

Chapter 8

Dynamic Link from Liftoff to Final Orbital Insertion for a MEO Space Vehicle

Jack K. Kreng and Gleason Q. Chen

Abstract

During the entire launch sequences from liftoff to final orbital insertion of a space vehicle (SV), adequate link requirements are to be maintained for telemetry, tracking, and command (TT&C) for uplink and downlink services, from launch vehicle (LV) and SV to ground stations (GS). A successful space vehicle launch required adequate link coverage with good radio frequency (RF) performance. The chapter is an extension of the IEEE/Aerospace Conference 2019 paper entitled Dynamic Link Analysis and Application for a MEO Space Vehicle published by the authors. The emphasis in this chapter is on the addition of the three distinctively different tracking waveforms and their associated links, used from liftoff to final orbital insertion. This chapter will describe the three required dynamic link analyses (DLA) to cover (a) the LV link from liftoff to its end of line of sight (LOS), (b) the LV link from LOS to Tracking and Data Relay Satellite System (TDRSS) at beyond line of sight (BLOS), and (c) the final tracking link using Space-to-Ground Link Subsystem (SGLS) or non-SGLS (NSGLS) link for the earliest or best separation time of the SV from the LV. The chapter discusses the concept of the dynamic link analysis, SV antenna switching schedule, recommended SV separation time, as well as the performance for different launch scenarios within the 24-h launch window. Topics include antenna patterns, launch trajectories, elevation angle and clock and cone angle geometry, and dynamic link budget.

Keywords: wide band communications, satellite tracking

1. Introduction

Recently, there was an interest to extend the present dynamic link analyses (DLA) beyond the early launch period to cover the period after the space vehicle (SV) separation from launch vehicle (LV), which includes both booster and second stage engine. The dynamic link from liftoff to final orbital insertion considers both geometric (visibility coverage) and radiometric (link margins for all downlink and uplink services) adequacy in the three launch stages. The purpose of the dynamic link study for the launch is to provide the earliest and accurate time for final SV separation and orbital insertion as compared to previous method which only relied on visibility tracking coverage to the end of line of sight (LOS).

The present DLA typically covers only two stages of LV tracking, including (a) liftoff to the end of LOS link and (b) the end of LOS to a period before SV payload (PL) separation from LV, using LV, to Tracking and Data Relay Satellite System (TDRSS) [1] satellite link, which is also called beyond line of sight (BLOS) link. A third SV tracking, after SV payload separation from LV, is a tracking link between SV and a ground station (GS). This third SV tracking is now added in this chapter.

The tracking link used from liftoff to the end of LOS uses a UHF noncoherent FSK signal for command and a digital FM or BPSK for tracking telemetry link as described in detail in [2]. From the end of LOS to BLOS, the tracking telemetry link is usually a BPSK or QPSK signal, using a NASA Tracking Data Relay Satellite System to relay tracking data from the LV to White Sands or Goddard ground station (WSGT/GRGT) and finally routing it to other user ground stations. After SV payload separation and orbital insertion, the SV tracking link to an Air Force Satellite Control Network (AFSCN) ground station [3] will use Space-to-Ground Link Subsystem (SGLS) or a non-SGLS (NSGLS) waveform described in [3, 4] and in Section 3 for tracking signals along the trajectory. In the following pages, supporting link analyses for the two LV and one SV tracking stages will be presented.

2. Geometry and coordinate definition for dynamic link analysis

2.1 Antenna coordinate system space vectors of interest

Figure 1 illustrates the antenna coordinate system space vectors of interest [5].

2.2 Antenna coordinate system definition

Figure 2 defines the antenna coordinate system used in this chapter. The azimuth (AZ or Φ) or clock angle is used in the antenna cut configuration. The elevation angle (EL or Θ) also called as cone angle is also used in the antenna gain data file. The antenna gain is a changing variable as a function of mission elapsed time (MET). The antenna gains are used in the following dynamic links.

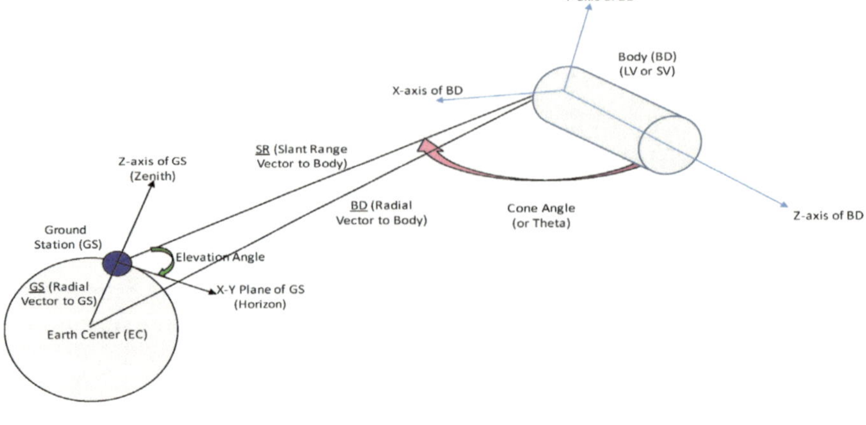

Figure 1.
Space vehicle and ground station vector definition.

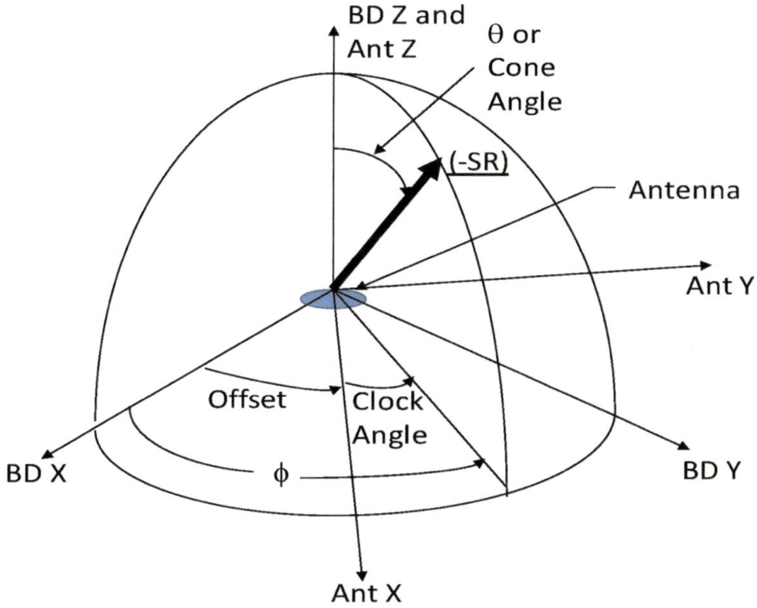

Figure 2.
Antenna coordinate definition.

3. Dynamic link calculation

3.1 Dynamic link formulas of interest

This section provides a summary of the dynamic link model of interest [6]. More detailed derivation of other variables, especially UVZBD, UVYBD, and UVXBD, can be found in [5]. For station elevation angle, either LV or SV elevation or cone angle (EL or theta or Θ), and LV or SV clock or azimuth angle (AZ or Phi or Φ), we have[1]

$$\text{Cone Angle} = \frac{180^0}{\pi}\cos^{-1}\left(\frac{\underline{\text{UVZBD}}\cdot(-\underline{\text{SR}})}{|\underline{\text{SR}}|}\right)\text{ in Deg} \tag{1}$$

$$\text{EL} = 90^0 - \frac{180^0}{\pi}\cos^{-1}\left(\frac{\underline{\text{GS}}\cdot\underline{\text{SR}}}{|\underline{\text{GS}}| * |\underline{\text{SR}}|}\right)\text{ in Deg} \tag{2}$$

$$\varphi = \frac{180^0}{\pi}\tan 2^{-1}\left(\frac{\underline{\text{UVXBD}}\cdot(-\underline{\text{SR}})}{|\underline{\text{SR}}|}, \frac{\underline{\text{UVYBD}}\cdot(-\underline{\text{SR}})}{|\underline{\text{SR}}|}\right)\text{ in Deg} \tag{3}$$

$$\text{Clock Angle} = \Phi - \text{Offset}$$

3.2 Tracking signals and link analyses along the trajectory

As mentioned in the introduction, there are three separate tracking signals along the flight trajectory that we need to analyze ensuring that they have adequate link margins of three dB or more. Most of the present DLA covers only two stages of launch coverage and neglecting the third stage coverage. The requirement for third stage tracking is explained below:

---

[1] The authors would like to thank Dr. James Yoh for his derivation of these formulas.

1. From liftoff to the end of LOS, the waveform for this link is generally a digital FM or BPSK for telemetry downlink as defined in the Range Commander Council (RCC) 119-88 [2]. The liftoff to LOS 5 link margin plot is shown in the left half of **Figure 3** for five different ground stations.

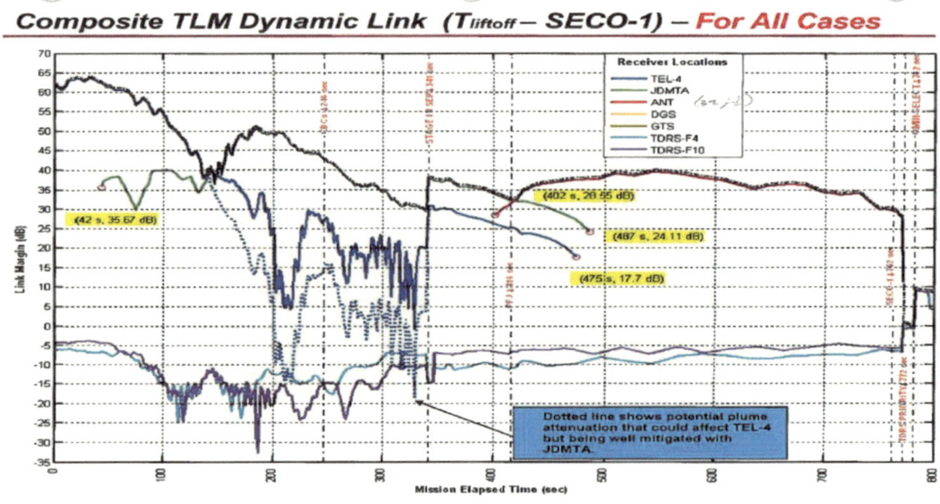

Figure 3.
TLM dynamic link margin from TEL4 to TDRSS versus mission elapsed time.

LV to TDRSS Range TLM Link				
Service:	SA		Mode:	
Frequency: (2200-2300MHz Typ.)	2211	MHz	Polarization:	LHCP
			Range (km):	40347
Space-Space Link (EELV)		Value	Units	Notes
1	EELV Transmit Power	44.15	dBm	User Provided Data
2	A/B Cable losses	1.5	dB	User Supplied Data
3	A/B Antenna Gain	2.1	dBic	User Supplied Data
4	Pointing loss	0.0	dB	User Supplied Data
5	EELV EIRP	44.7	dBm	(1)-(2)+(3)-(4)
6	Space Loss	191.45	dB	LMAC SR= 132 339 260 ft ,near Geo
7	Atmospheric Loss	0.00	dB	Not considered
8	Multipath Loss	0.00	dB	Not considered
9	Polarization Loss	0.10	dB	Not considered
10	Pisotropic at Input to TDRSS	-146.85	dBmi	Piso = (5)-(6)-(7)-(8)-(9)
11	TDRS SA Antenna G/T	10.4	dB/K	CLASS Database
12	Self/Mutual Interference (Lim) Loss	0.00	dB	CLASS Database
13	C/No @ TDRSS	62.15	dB-Hz	C/No = Piso+G/T+1/k-Lim k = -198.6dBm/K-Hz
14	Bandwidth BW	54.1	dB-Hz	256 KBPS BPSK, Detection BW 256KHz
15	C/N =SNR @ TDRSS	8.05	dB	C/No - BW (13-14)
Space-Ground Link (NASA Data)				
16	TDRSS EIRP	26.33	dBW	CLASS Database
17	Path Loss	207.32	dB	CLASS Database
18	Atmospheric Loss	0.25	dB	CLASS Database
19	Polarization Loss	0.03	dB	CLASS Database
20	Rain Attenuation	6.00	dB	99.5%WSC,~97%Guam
21	Pisotropic @ WSC Station	-187.27	dBW	Piso =(16)-(17)-(18)-(19)-(20)
22	Ground G/T	41.00	dB/K	CLASS Database
23	TDRS Dwnlnk C/No (Thermal)	82.33	dB-Hz	C/No= Piso @Gnd+Gnd G/T+ 1/k
24	IM/XPOL degradation	3.94	dB	Class analysis (P/IM=20.83dB)
25	TDRSS Dwnlnk C/No (Total)	78.39	dB-Hz	C/No-degradation (23)-(24)
26	Bandwidth BW	54.1	dB-Hz	256KBPS BPSK, Detection BW 256KHz
27	TDRSS Dwnlnk C/N =SNR (Total)	24.29	dB	C/No - BW (25-26)
Ground Terminal				
28	C/N @ Ground	7.95	dB	(15) // (27)
29	Bandwidth BW	54.1	dB-Hz	256 KBPS BPSK, Detection BW 256KHz
30	MA Net combiner Gain	0.00	dB	MA Ant. Not Used, SA Only
31	C/No @ Ground	62.05	dB-Hz	C/N+BW+Gain
		I-Ch	Q-Ch	
32	Channel power Split	-3.01	-3.01	User Provided Reference Data
33	Ch C/No @ Ground	59.04	NA	dB-Hz (31)+(32)
34	Bit Rate	54.1	NA	dB-BPS, user provided, 256KBPS
35	Eb/No =SNR into Demodulator	4.94	NA	dB, (33)-(34)
36	Dynamics loss	0.00	NA	dB, Not considered
37	User constraint loss	0.00	NA	dB, WSC
38	RFI loss	0.1	NA	dB, WSC
39	Implementation loss	1.0	NA	dB, WSC
40	Net Eb/No	3.84	NA	dB, (35)-(37)-(38)-(39)
41	Theoretical Req. Eb/No	4.20	NA	dB, BER=1e-5 with FEC, WSC
42	Margin	-0.36	NA	dB, Margin=Net Eb/No – Req Eb/No
43	Margin (W/o Ch. Pwr Split)	+2.65	NA	dB, (42)- (32) No Pwr Split Loss

Table 1.
LV to TDRSS range TLM link (based on NASA source).

2. From the end of LOS to NASA TDRSS at geosynchronous orbit, the telemetry link is a BPSK or QPSK (telemetry + data) which is sent from LV to TDRSS to be relayed to White Sands or Goddard ground station (WSGT/GRGT), as shown in the second right half of the **Figure 3** and in more link details in **Table 1**.

3. For the third link after SV payload separation, when the satellite or SV starts its transfer orbit, the tracking link from the SV payload (or bus) to an AFSCN ground station will be in SGLS, Unified S-Band (USB), or a NSGLS waveform as described in more detail in [3, 4]. For a more secure tracking, SV normally will be using SGLS link for tracking as described in [4], with a MEO satellite in **Table 2** as an example. A commercial and less secure SV launch may use USB

	Parameter	Unit	Uplink Link (Ground Station to SV)	Downlink Link (SV to Ground Station)
A	B	C	D	F
	Transmit System			
5	Carrier frequency (fc)	MHz	1791.700	2237.500
6	Power (Pt)	dBm	57.00	37.00
7	Tx circuit loss	dB	-1.00	-3.00
8	Antenna diameter	m	10.0000	NA
9	Antenna peak gain (Gt)	dBi	43.94	2.00
10	EIRP	dBm	99.94	36.00
	Transmission Media			
13	Atmospheric loss	dB	-0.10	-0.20
14	Polarization Loss	dB	-1.80	-0.30
15	Total transmit losses	dB	-1.90	0.50
16	Slant range/Max path	km	24713.00	24713.00
17	Space loss (Ls)	dB	-185.37	-187.30
18	Received isotropic power (RSSi, Piso)	dBmi	-87.32	-151.80
	Receive System			
21	Antenna diameter	m	0.058	10.060
22	Polarization loss	dB	-0.20	-0.20
23	Antenna gain-peak (Gr)	dBi	2.00	45.23
24	Received power @ antenna hub (Pr=C)	dBm	-85.52	-106.77
25	Antenna temperature (T_A)	°K	290.00	22.00
26	Receiver line loss (L)	dB	10.00	1.00
27	Receiver noise figure (NF)	dB	2.50	1.10
28	System temperature (Ts)	°K	5157.01	202.32
29	Antenna gain/System temp. (Gr/Ts)	dB/K	-35.12	22.17
30	Noise spectral density @ ant. hub (No)	dBm/Hz	-161.48	-175.54
31	C/No at Receiver Input	dB-Hz	75.95	68.77
	Carrier Service		**UL LINK (to SC)**	**DL LINK (to Station)**
34	Carrier modulation loss	dB	-2.25	-7.45
35	Residual Carrier C/No @ rcvr input	dB-Hz	73.70	61.32
36	Loop Noise bandwidth, B	Hz	20.00	20.00
37	Received C/N	dB	60.69	48.31
38	Required C/N	dB	15.00	17.00
39	Service Margin	dB	45.69	31.31
	CMD Service (3-FSK)		**UL LINK (to SC)**	**DL LINK (to Station)**
43	Command modulation index	rad	0.90	1.12
44	Command modulation loss	dB	-5.22	-7.86
45	Rcvd CMD (Pcmd/No) @ rcvr input	dB-Hz	70.73	60.91
46	Data rate (Rb)	bps	1000.00	
48	Received Eb/No	dB	40.73	
49	Required Eb/No	dB	17.60	
50	Service Margin	dB	23.13	
	PRN Ranging (coherent only)		**UL LINK (to SC)**	**DL LINK (to Station)**
53	Ranging modulation index	rad	0.30	0.37
54	Ranging modulation loss	dB	-12.45	-18.84
55	Rcvd Ranging (Prn/No) @ rcvr input	dB-Hz	63.51	49.94
56	Data Rate, Noise Bandwidth, B	Hz	NA	10.00
58	Received C/N	dB	NA	39.94
59	Required C/N	dB	NA	28.00
60	Service Margin	dB	NA	11.94
	Telemetry Service (BPSK on 1.024 MHz S/C)		**UL LINK (to SV)**	**DL LINK (to Station)**
64	TLM modulation index	rad	N/A	1.30
65	TLM modulation loss	dB	N/A	-5.94
66	Rcvd TLM (Ptlm/No) @ rcvr input	dB-Hz	N/A	62.83
67	Data rate (Rb)	kbps	N/A	1.00
69	FEC Coding Gain	dB	N/A	0.00
70	Received Eb/No	dB	NA	32.83
71	Required Eb/No	dB	N/A	9.60
72	Service Margin	dB	N/A	21.23

Table 2.
Link budgets for SGLS TT&C uplink and downlink services [4].

	Parameter	Unit	Uplink Link (Ground Station to SV)	Downlink Link (SV to Ground Station)
A	B	C	D	F
	Transmit System			
5	Carrier frequency (fc)	MHz	1791.700	2237.500
6	Power (Pt)	dBm	57.00	37.00
7	Tx circuit loss	dB	-1.00	-3.00
8	Antenna diameter	m	10.0000	NA
9	Antenna peak gain (Gt)	dBi	43.94	2.00
10	EIRP	dBm	99.94	36.00
	Transmission Media			
13	Atmospheric loss	dB	-0.10	-0.20
14	Polarization Loss	dB	-1.80	-0.30
15	Total transmit losses	dB	-1.90	-0.50
16	Slant range/Max path	km	24713.00	24713.00
17	Space loss (Ls)	dB	-185.37	-187.30
18	Received isotropic power (RSSi, Piso)	dBmi	-87.32	-151.80
	Receive System			
21	Antenna diameter	m	0.058	10.060
22	Polarization loss	dB	-0.20	-0.20
23	Antenna gain-peak (Gr)	dBi	2.00	45.23
24	Received power @ antenna hub (Pr=C)	dBm	-85.52	-106.77
25	Antenna temperature (T_A)	°K	290.00	22.00
26	Receiver line loss (L)	dB	10.00	1.00
27	Receiver noise figure (NF)	dB	2.50	1.10
28	System temperature (Ts)	°K	5157.01	202.32
29	Antenna gain/System temp. (Gr/Ts)	dB/K	-35.12	22.17
30	Noise spectral density @ ant. hub (No)	dBm/Hz	-161.48	-175.54
31	C/No at Receiver Input	dB-Hz	75.95	68.77

Table 3.
Typical C/No for uplink and downlink budgets.

or NSGLS for SV tracking instead of using SGLS waveform. **Table 3** shows link budget for uplink and downlink C/No example for tracking links 1 and 2. **Table 2** shows SGLS telemetry, tracking, and command (TT&C) link budget for tracking link 3 for a MEO satellite. If the SV is using a USB or a NSGLS [3], the tracking waveform can be an AQPSK signal with telemetry on the I channel and ranging on the Q channel.

4. Basic link parameters and formulas

This section describes the basic link parameters including LV or SV transmitter power amplifier gain (Pt), transmitter antenna gain (Gt), space loss (Ls), received isotropic power (RIP), and received (C/No = SNR).

A modulation signal or information data is generated at a ground station, in an LV or in an SV. This modulation signal will be used to modulate onto the radio frequency (RF) carrier to become a modulated transmit signal. This transmit system will be consisting of a high-power amplifier (HPA) which amplifies the signal to generate an output power expressed in dBW (conversion from Watts to dBW is simply dBW = 10*log10(Watts)); some cables and circuits with a loss and an antenna with a gain are added together as shown below. The output from the transmit system is therefore an effective isotropic radiated power or EIRP, which can be found in either the uplink or the downlink of an LV or an SV tracking system.

$$\text{EIRP} = P_T + L_C + G_t \text{ in dBW} \tag{4}$$

where G_t = transmit antenna gain, in dBi; L_C = transmit circuit loss, in negative dB; and P_T = HPA output power, in dBW.

For the uplink, where the transmit antenna is located on the ground, the antenna can be easily directed to an LV or SV. This transmit antenna is likely to be a directional high gain dish antenna for connectivity with an LV or SV located possibly far away. In **Table 3**, line 9, a typical ground station has a large parabolic antenna dish with a gain Gt = 43.94 dBi using the following formula.

$$G_t = 10 * \log 10\left[\eta * \left(\frac{\pi * f_C * D}{c}\right)\right] \text{ in dBi} \quad (5)$$

where η = average antenna efficiency (0.70 in the calculation in **Table 3**); f_C = uplink frequency, in Hz; D = antenna diameter, in m; and c = speed of light, in m/s.

For the downlink, the transmit antenna is typically an omnidirectional antenna that covers a larger portion of the sky or the Earth, in which case the antenna gain is small (e.g., 2 dBi) and is either specified as in line 9 of **Table 3** or can be interpolated from values extracted from a table of antenna gain pattern with a specific AZ, EL, and MET, using a mission specific launch trajectory.

In general, there are terms that may be added to **Tables 2** and **3**. For example, the uplink transmit antenna in **Table 3** may have two more loss terms, namely, radome loss to account for any loss for a radome and pointing loss to account for any pointing error in directing the boresight of the antenna. For **Table 3** they are both negligible and ignored except for the polarization losses. The transmit and receive polarization losses (lines 14 and 22, respectively, in **Table 3**) can be accounted for as one-single combined receive polarization loss.

4.1 Transmission medium and losses

The signal path traverses through the transmission medium, in between transmit and receive systems. When the distance between transmit and receive systems increases, the signal beam has an angular spread which decreases the signal power collected by a receiving antenna. We know that the portion of the transmission medium near the ground station depends on the Earth's atmosphere which attenuates the signal to different degrees, dependents on the frequency, the altitude of the GS, and the angle of the signal path through the atmospheric (GS elevation angle). Beyond the Earth's atmosphere, the signal path traverses through the space with little atmospheric attenuation, only with free space loss to account for. Therefore, there are essentially two losses through the transmission medium, namely, the space loss to account for the spreading of the signal beam and the additional atmospheric loss [7].

$$L_S = 10 * \log 10\left[\left(\frac{c}{4\pi * f_C * SR}\right)^2\right] \text{ in dB} \quad (6)$$

where f_C = carrier frequency, in Hz; c = velocity of light, in m/s; and SR = slant range between GS and SV, in m.

At L and S bands, the atmospheric loss is very small, at less than 0.001 dB/Km or 0.1 dB/100Km for a link availability of better than 98% [8].

4.2 Received isotropic power

The transmit signal, after accounting for the space and atmospheric losses, and its signal path terminates at the antenna of either the ground station, the SV, or the LV receiving system. Before considering the characteristics of the receive antenna and the receiver, a good indication of the signal strength is given by the received isotropic power. RIP is simply the transmitter EIRP after subtracting off the losses of the transmission medium, i.e.

$$\text{RIP} = \text{EIRP} + L_S + L_A \text{ in dBW} \tag{7}$$

where L_A is the atmospheric loss extracted from tables or curves, in negative dB (very small at 0.02 dB/Km per Datron chart in L and S bands). Ls can also be obtained from the Datron calculator [8].

4.3 Receive system and performance

The last portion of **Table 3** addresses the receiving system and assesses how well it performs. This section involves with the calculation of the signal strength and the noise strength, resulting in the ratio of signal power over noise power density (C/N_0). In general, we first address the signal power and then the noise power density. Line 21 provides for the receive antenna size consistent with the receive antenna gain to be calculated later. The next parameter in **Table 3** is the polarization loss, which accounts for the mismatching between the polarization axial ratios of the received signal and the receiving system. The axial ratio is the ratio of the major axis of an ellipse to its minor axis. For circularly polarized signal, the ratio should be 0 dB. Any deviation from 0 dB results in a polarization loss. Line 23 shows the values of receive antenna gain. For downlink in which the receive antenna is a dish antenna located at GS, G_r is calculated using the standard dish antenna equation (similar to Eq. (6) for G_t).

$$G_r = 10 * \log 10 \left[\eta * \left(\frac{\pi * f_C * D}{c} \right) \right] \text{ in dBi} \tag{8}$$

where η = average antenna efficiency (assumed to be 0.6 in the calculation in **Table 3**); f_C = downlink frequency, in Hz; D = antenna diameter, in m; and c = velocity of light, in m/s.

At the end, the received power at the antenna feed is just the sum of RIP, minus the polarization loss, plus the receive antenna gain, i.e.

$$C = \text{RIP} + L_P + G_r \text{ in dBW} \tag{9}$$

where L_P is the polarization loss, in negative dB.

For the downlink transmit antenna on the SV, as in the case of uplink receive antenna, the SV antenna is a broad beam Earth coverage (EC) omnidirectional type of SV antenna, with a gain of 2 dBi (see line 23 of **Table 3**).

For the noise power density (N_0), we need to calculate the system temperature (T_S) measured at the antenna feed. The system temperature is the sum of antenna sky temperature (T_A) and the composite temperature from antenna line loss (L_L) and low noise amplifier noise figure (NF) which are referred to the antenna feed. In linear quantity, T_S is given by [1].

$$T_S = T_A + \left(10^{(L_L+\text{NF})/10} - 1\right) * 290 \text{ in Deg} - \text{K} \tag{10}$$

where NF = low noise amplifier noise factor, in dB and L_L = line loss, in dB.

The noise density (N_0) is given by

$$N_0 = k\bar{dB} + 10 * \log(T_S) \text{ in dBW/Hz} \tag{11}$$

where k = Boltzmann's constant in dB = −228.6 dBW/KHz.

Using Eqs. (9) and (11), the ratio C/N_0 at the receiver input is obtained in **Table 3**, line 31. It represents the final product before going into specific service(s) such as telemetry and ranging to evaluate their performance.

5. Space vehicle link services

For many SVs, we are interested in their uplink and downlink services. **Table 2** shows an example, taken from an IEEE paper [4]. This is the standard link budget, where the ground station is an AFSCN [3] remote tracking station (RTS) using SGLS waveform [3, 4]. The waveform is described in an AFSCN interface control document (ICD) [3] and is implemented in DLA, although other waveforms can be readily incorporated. The uplink has two services of interest—carrier and command—while the downlink has three services of interest: carrier, ranging, and telemetry. In general service margins are calculated for these five services. For the SGLS waveform, command is coupled with ranging and modulated on the uplink carrier; therefore command is also turned around at the SV along with ranging. This SGLS turnaround process explains the reason that **Table 2** shows a power allocation for command in the downlink and no calculation for its margin. As a result, downlink power allocated to command is essentially wasted while robbing power from other downlink services. The requirements and service margins for command and telemetry are expressed in E_b/N_0, since it is the bit error rate (BER) that counts for both cases. The carrier and ranging are expressed in C/No given by a specific station. For ranging, it is the autocorrelation value between the decoded ranging code and the transmitted ranging code that needs to be maximized in order to successfully perform accurate ranging.

Table 2 represents uplink and downlink budgets for SGLS TT&C. Let us address the important aspects of the calculation of uplink and downlink services in the next few subsections. The role of modulation indices is to divide up the power for allocation to services. The modulation index is expressed in radians so that it can go right in as an argument in a sinusoidal or Bessel expression. If the modulation indices of all services are zero radians, no power is allocated to the services, and the carrier retains all the link power calculated in Section 4. If the modulation indices of services are not zero, portions of the power are taken from the carrier and allocated to the services. The remaining power stays with the carrier as the "residual carrier power."

5.1 SNR and link margin calculation

After SV separation, we are dealing with the SV uplink and downlink using SGLS or NASA Unified S-Band waveforms as described in [3, 4]. For telemetry service, the requirement is SNR = Pservice/NoB = Eb Rb/NoB = Eb/No in dB. For carrier and ranging, the requirements are stated in terms of C/No as mentioned before. For acquisition, the uplink carrier loop bandwidth could be as high as +/− 100 KHz, while its tracking bandwidth could be as small as a few Hz. For the station the carrier tracking loop bandwidth is about 20–50 Hz, as in **Table 2** in line 36. For ranging, the bandwidth of 10 Hz represents ranging tracking loop bandwidth (**Table 2**, line 56), which corresponds to the sampling rate of the autocorrelation

No.	Uplink modulation	Carrier	Ranging	Command
Uplink service modulation losses for SGLS and NASA USB				
1	AM-3FSK/PRN RNG/PM (SGLS unfiltered uplink) [3, 4]	$J_o^2(\beta_1).cos^2(\beta_2)$	$J_o^2(\beta_1).sin^2(\beta_2)$	$2J_1^2(\beta_1).cos^2(\beta_2)$
2	BPSK/PRN RNG/PM (USB filtered case) (Eqs. (1-22) and (1-23)) [9]	$J_o^2(\beta_1)cos^2(\beta_2)$	$J_o^2(\beta_1)\ sin^2(\beta_2)$	$2J_1^2(\beta_1)\ cos^2(\beta_2)$
3	BPSK/Tone RNG/PM (USB unfiltered case) (Eqs. (1-18) and (1-21)) [9]	$J_o^2(\beta_1)\ J_o^2(\beta_2)$	$2J_o^2(\beta_1)J_1^2(\beta_2)$ + ...	$2J_o^2(\beta_2)\ J_1^2(\beta_1)$ + ...
Downlink modulation losses for SGLS and NASA USB				
No.	**Downlink modulation**	**Carrier**	**Ranging**	**Telemetry**
4	PRN RNG/PSK TLM/PM (SGLS filtered case) (Eqs. (2-18) and (2-20)) [4]	$J_o^2(\beta_1')J_o^2(\beta_2')$ $J_o^2(\beta_3)$	$2J_o^2(\beta_1')\ J_1^2(\beta_2')$ $J_o^2(\beta_3)$	$2J_o^2(\beta_1')J_o^2(\beta_2')$ $J_1^2(\beta_3)$
5	Tone RNG/PSK TLM/PM (USB filtered case) (Eqs. (2-18) and (2-21)) [9]	$J_o^2(\beta_3)J_o^2(\beta_2')$	$2J_o^2(\beta_3)J_1^2(\beta_2')$	$2J_o^2(\beta_2')J_1^2(\beta_3)$

Table 4.
Uplink and downlink service modulation losses for SGLS and NASA USB.

value between the detected ranging code and the transmitted ranging code. For command and telemetry, the requirements are expressed in terms of E_b/N_O. The command and telemetry data bit rates of 1000 bps each are representing the lower end of their SGLS choices. As shown in **Table 2**, the results from the SNR calculation are the values of C/N and E_b/N_O for various received uplink services (lines 37 and 48 for command C/N or Eb/No) and for various downlink services (line 55 for ranging service Prng/No and line 70 for telemetry service Eb/No). The uplink service modulation losses for SGLS and NASA USB with subcarrier (S/C) [3, 4, 9] are shown in **Table 4**. The downlink service modulation losses for SGLS and NASA USB with subcarrier (S/C) [3, 4, 9] are shown in **Table 4**. Also note that β_1, β_2, and β_3 represent the modulation indices for command (CMD), ranging (RNG), and telemetry (TLM), respectively, per [3, 4, 9]. These uplink and downlink modulation losses are in lines 34, 44, and 54 in **Table 2**.

For NSGLS waveforms such as the direct mod BPSK, QPSK, and AQPSK, the service mod losses are negligible. Finally, the calculated service SNR in **Table 2** is compared with the required SNR to obtain the link margin for each service. The required SNR values capture all the performance requirements for the services, such as ranging accuracy, tracking loop loss likelihood, bit error rate, and others.

6. Launch vehicle dynamic link

Before SV separation from the LV, we are also interested in the dynamic link from a ground station to the LV, from liftoff to the SV after separation, along the entire LV flight path using the tracking stations in the line of sight (TEL4, JDMTA, ANT, DGS, TDRSS). The waveforms for this LV tracking are described in Range Commander Council (RCC) handbook [2]. One must ensure that the downlink telemetry link from the launch vehicle to these ground stations and TDRSS relay satellite are positive as can be seen in **Figure 3**. The basic LV range modulations are digital FM, BPSK, QPSK, AQPSK, etc. as discussed in RCC [2]. In **Figure 4**, dynamic LV slant range “received TLM Eb/No” and “TLM link margin” for a specific mission

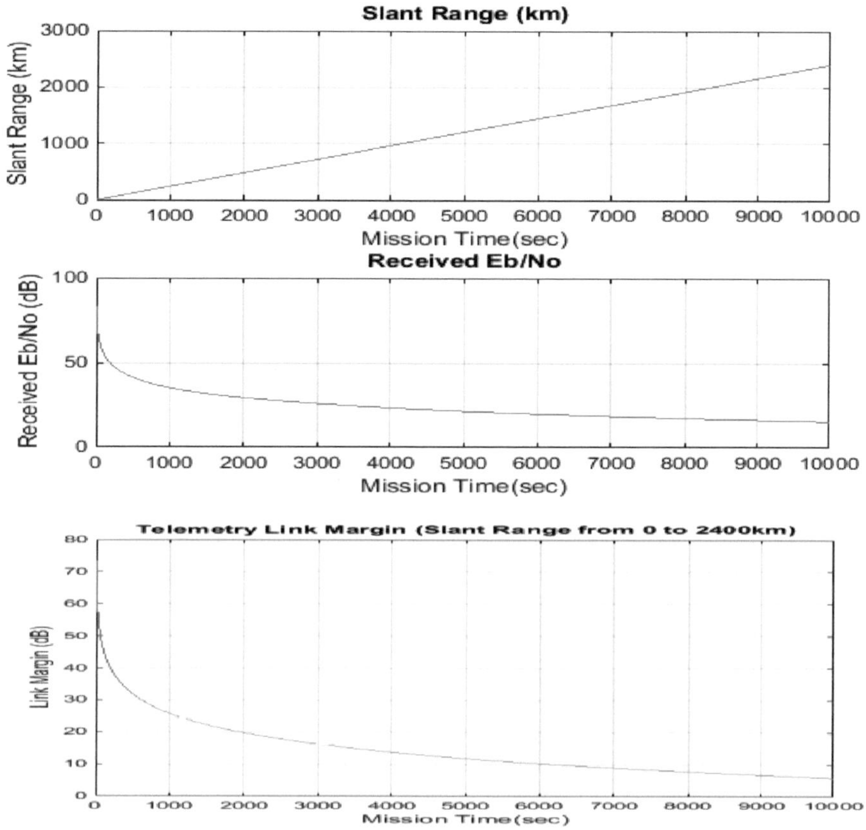

Figure 4.
LV slant range received TLM Eb/No and TLM link margin.

are displayed together. As an example a specific LV to TDRSS BPSK link using NASA data is shown in **Table 1**.

7. Conclusion

This chapter discusses the required three DLAs and related tracking waveforms to cover the three launch stages, namely, (a) the launch vehicle tracking link from liftoff to its end LOS using the digital FM or BPSK signal, (b) the launch vehicle tracking link from LOS to TDRSS at BLOS using NASA USB signal, and (c) the final tracking link from SV to an AFSCN ground station using AFSCN SGLS, AFSCN NSGLS, or NASA USB waveforms. In the third tracking link case, BPSK, QPSK, or AQPSK waveforms were used, in which for QPSK and AQPSK, the telemetry data is put on the I channel and the ranging signal is on the Q channel.

The chapter shows that good telemetry link margins from LV to tracking stations such as TEL4, JDMTA, and ANT or to a NASA TDRSS relay satellite can be achieved using digital FM, BPSK, QPSK, or AQPSK signals, after SV separation. The chapter also shows that good tracking link margins can be achieved from SV to AFSCN ground stations, including IOS or DGS as the first contact station.

Author details

Jack K. Kreng* and Gleason Q. Chen
Aerospace Corporation, El Segundo, CA, USA

*Address all correspondence to: jack.k.kreng@aero.org

References

[1] Sklar B. Digital Communications Fundamentals and Applications. New Jersey: Prentice Hall; 1988. p. 221

[2] RCC, Range Commander Council, Telemetry Group. Document 119-88; 1988. (approved for public release)

[3] Air Force Satellite Control Network (AFSCN) Interface Control Document: Range Segment to Space Vehicle, ICD-000502A, prepared by The Aerospace Corporation for Space and Missile Systems Center; 2011

[4] Kreng JK, Ardeshiri MM. Effects of turnaround command on SGLS C/No and SNR performance. In: 2018 IEEE Aerospace Conference, Paper #2513; 2018

[5] Krikorian YY et al. Dynamic link analysis tool for a telemetry downlink system. In: 2004 IEEE Aerospace Conference Proceedings; 2004. pp. 3994-4001

[6] Spiegel MR, Lipschutz S, Liu J. Mathematical Handbook of Formulas and Tables, Third Edition, Section 10. New York: McGraw Hill; 2009

[7] Kreng J, Do S, Mathur A. USB Command Link Via TDRSS to Satellites and Possibly Space Launch Vehicles (LV). AIAA; 2008

[8] Datron/Transco Inc., USA. Antenna, Space and Atmospheric Calculator

[9] Kreng J, Krikorian Y, Raghavan S. Telemetry, tracking, and command link performance using USB/STDN waveform. IEEE/Aerospace Journal. 2007

Chapter 9

Effective Algorithms for Detection Outliers and Cycle Slip Repair in GNSS Data Measurements

Igor V. Bezmenov

Abstract

The chapter describes effective algorithms that are often used in processing data measurements in Global Navigation Satellite Systems (GNSSs). Existing effective algorithm was developed for detection and elimination of outliers from GNSS data measurements. It is based on searching for a so-called optimal solution for which standard deviation and maximum absolute deviation of the measured data from mean values do not exceed specified threshold values, and the number of the detected outliers is minimal. A modification of this algorithm with complexity of $N \log_2 N$ is discussed. Generalization of the existing algorithm to the case when data series included some unknown trend will be presented. The processing trend is assumed to be described by an unknown function of time. The generalized algorithm includes the outlier detection algorithm and trend searching algorithm that has been tested using simulated data. A new algorithm will be presented for cycle slip repair using Melbourne-Wübbena linear combination formed from GNSS data measurements on two carrier frequencies. Test results for repair data in the case of multiple (cascade) cycle slips in actual observation data will also be presented in this chapter.

Keywords: global navigational satellite systems (GNSSs), GNSS measurements, outliers, data screening, optimal solution, trend function, Melbourne-Wübbena combination, cycle slips

1. Introduction

At present, five constellations of GNSS satellites are involved in the formation of observational data, which serve as a source for many applications related to navigation, geodesy, geodynamics, and in the performance of solving of many fundamental problems. These are American Global Positioning System (GPS), Russian Global Navigation Satellite System (GLONASS), European Galileo, Chinese BeiDou, and Japanese Quasi-Zenith Satellite System (QZSS). The satellites of each of the operating systems transmit signals, as a rule, on two L-band carriers, which are received by GNSS receivers. A large number of stations equipped with GNSS receivers and located around the World are part of the International GNSS Service (IGS) network. These stations generate observation data files and transmit them to international databases in real time [1, 2], after which these data become available for use by many institutions and laboratories over the World. When solving applications, the measurement data go through various processing steps. Significant element of the data processing is the detection of rough measurements and removal them from the further processing. Despite the fact that many of the laboratories use a high-end application of the software regarding accuracy, reliability, and robustness, the presence of rough measurements in the observational data excludes the possibility of obtaining an accurate final result. In order to obtain results of unprecedented accuracy, the measurement data must be cleared of coarse measurements or outliers. It should be noted that the concept of outliers is key in the measurement processing theory [3], and there is no general definition for it. In order to distinguish outliers from the rest of the measured data, in some cases, the deviation of the data series values from some average value of the data is considered. If the deviation from the average is exceeded by a predetermined threshold value, the measured value is considered as an outlier. Such an approach has a significant disadvantage that the exact mean is generally unknown, and the estimate obtained by averaging a series may be very inaccurate due to outliers. Existing iterative procedures are also based on the idea of calculating deviation from the average and often result in the unjustified rejection of many observations. Reducing the data involved in processing may, in turn, result in a loss of accuracy of the final result.

This chapter describes the outliers cleaning algorithm for GNSS data. The proposed algorithms are based on the search for the so-called optimal solution with the minimum amount of invalidly rejected data. The algorithm for accelerated detection of outliers in a large amount of measurements has been developed, as well as an algorithm for detecting outliers in data containing an unknown trend. In conclusion, the algorithm of jump detection in the Melbourne-Wübbena combination [3–5], including the developed procedure of cleaning data from outliers, is considered.

In Section 2, the problem of searching for the so-called optimal solution is formulated. Section 3 provides a search algorithm, with a common number of arithmetic operations not exceeding $\sim N^2$. Section 4 presents the test results for actual measurements in global navigational satellite systems at two carrier frequencies. The searching of outliers was performed in the Melbourne-Wübbena combination. In Section 5, the assertions that are the mathematical prerequisites for justifying a fast outlier search algorithm are proved. In Section 6, the fast outlier detection algorithm with the number of arithmetic operations of $N \log_2 N$ is proposed. Section 7 describes the case of data with unknown trend. Iterative procedure of outlier search is proposed based on the finding of suitable trend approximation in polynomials class. The idea of excluding coarse measurements is based on finding a so-called minimizing set of measurement data of a given length. This distinguishes the proposed algorithm from known similar procedures in which outliers are detected by exceeding a preset threshold. The test results with simulated data are given. Sections 8 and 9 discuss the problem of detecting jumps in the Melbourne-Wübbena combination. An algorithm is proposed that includes the outlier cleaning procedure based on the search for the optimal solution. Section 10 shows the numerical calculations with real data for algorithms presented in Sections 3–9. Section 11 concludes the chapter.

2. Statement of the problem: a Melbourne-Wübbena combination

Often, measurements y_j taken at moments of time j can be presented in the form:

$$y_j = f_j + z + \xi_j; j = 1 \ldots N, \tag{1}$$

where f_j is a trend function, depending on physical process and as a rule is unknown, the sum $z + \xi_j$ is unknown random variable imposed on the trend with unknown constant z and a centered random variable ξ_j.

Detection of outliers in data series expressed in Eq. (1) with unknown trend is uncertain since the concept of measurement or outliers itself is uncertain. In many cases, however, the trend function is known a priori.

For example, many data processing programs often use different linear combinations formed of code and phase measurement data to eliminate unknown parameters. One such combination is the Melbourne-Wübbena combination composed of both, carrier phase and code observables as described by Melbourne [4] and Wübbena [5]. This combination eliminates the effect of the ionosphere, the geometry, the clocks, and the troposphere [3], and it is often used to detect loss of carrier phase capture in the preprocessing stages. The Melbourne-Wübbena combination generated for a specific satellite-receiver pair can be presented in the form of the sum of three terms [6]. One of the terms includes the integer wide-lane ambiguity for the two carrier frequencies [3]; the second component accounts for the satellite and receiver instrumental delays; and the third component is the measurement noise, including carrier phase and code multipath. Thus, during a time interval where the integer wide-lane ambiguity does not change, the Melbourne-Wübbena combination can be written as formula (1) with $f_j = \text{const}$.

Another example is satellite clock correction values derived from navigation message data, which can also be represented as in Eq. (1) with $f_j = dj + a$, where d and a are the drift and offset parameters known from navigation data, respectively. In the case where the trend is known, the measurement data after the trend subtraction can be represented as, assuming N observations:

$$y_j = z + \xi_j; j = 1 \dots N, \tag{2}$$

with an unknown constant z, which we cannot be determined in advance, because the random value ξ_j is not known and may contain outliers.

In Sections 2–6, we consider the case where the trend is known a priori, that is, the data can be presented as Eq. (2). A problem with an unknown trend will be discussed in Section 7. In principle, the outlier detection procedure described below is not affected by the measurement format expressed in Eq. (1) or (2); it can be applied to any set of data measurements y_j.

The preliminary processing task includes rejection of rough measurements or outliers from data series (2). In other words, it is necessary to find a set $Y_L = \left\{y_{j_1}, \dots, y_{j_L}\right\}$ of L elements, where L is the length for which the following conditions are satisfied:

$$\sigma_{Y_L} = \sqrt{(L-1)^{-1} \sum_{j \in \{j_1, \dots, j_L\}} \left(y_j - z\right)^2} \le \sigma_{max}, \tag{3}$$

$$\left|y_j - z\right| \le 3 \cdot \sigma_{max}; y_j \in \left\{y_{j_1}, \dots, y_{j_L}\right\}, \tag{4}$$

$$L \ge \text{MINOBS}, \tag{5}$$

where σ_{Y_L} and σ_{max} are the standard deviation (SD) and their associated specified threshold values; MINOBS is a parameter limiting from below the length of the required set of measured values (e.g., 10), and we will assume hereafter that MINOBS $< N$.

The values y_j that are not included in the set Y_L are treated as coarse measurements and removed from further processing. Typically, expressions in Eqs. (3)–(5) are the only conditions considered in processing programs when screening out rough measurements. Below we will formulate the problem of searching for a solution, complementing the conditions expressed in Eqs. (3)–(5) with two extreme conditions [7].

1. First, we will require that the length of the set sought be the maximum, that is, the number of measurements deemed to be coarse is the minimum:

$$L \to \max. \tag{6}$$

Note that for the predetermined values y_j, the problem solution satisfying conditions in Eqs. (3)–(5) may not exist (e.g., when y_j includes a trend, in particular when y_j is an arithmetic progression with a step greater than σ_{max}). In the case when the solution does exist, we will denote the value L at which the maximum of Eq. (6) is reached as L_{max}. Note that the condition expressed in Eq. (6) does not ensure the uniqueness of the set because several sets of length L_{max} can be found that satisfy the conditions in Eqs. (3)–(5).

2. From all possible sets that satisfy conditions expressed in Eqs. (3)–(5) and (6), we will select the one for which the variable σ_{Y_L} receives the smallest value:

$$\sigma_{Y_L} \xrightarrow[Y_L=\left\{y_{j_1}, \dots, y_{j_L}\right\};\, L=L_{max}]{} \min. \tag{7}$$

Let us define Y_{opt} as follow:

Definition 1. *For a given sequence of values y_j, $j = 1, 2, \dots N$, the set of values:*

$$Y_{opt} = \left\{y_{j_1}, \dots, y_{j_{L_{max}}}\right\}, \tag{8}$$

satisfying conditions in Eqs. (3)–(7), we refer to as the optimal solution of the problem expressed in Eqs. (3)–(7). The corresponding SD value is denoted by σ_{opt}.

Thus, the problem consists in the creation of a search algorithm for the optimal solution of the problem shown in Eqs. (3)–(7).

In a practical situation, the precise value z, given conditions in Eqs. (3) and (4), is not known. We will estimate the values using the following formula:

$$z = L^{-1} \sum_{j \in \{j_1, \dots, j_L\}} y_j. \tag{9}$$

Note that the value z depends on the required solution, which will complicate its search.

Usually, iterative methods are used to find a solution to problem expressed in Eqs. (3)–(5). For example, the algorithm implemented in the observation data smoothing program (see [3]) is designed to find a set Y satisfying the conditions given by Eqs. (3)–(5). The proposed step-by-step algorithm is based on iterations (the index number of iteration is designated by the upper index in parentheses):

Step 1: Initialization: $Y^{(0)} = \{y_1, \dots, y_N\}$; $Level^{(0)} = 10^{20}$; k = 0.

Step 2: Checking the length of the set $Y^{(k)}$; if it is less than MINOBS, then the process comes to an end and a solution is not found.

Step 3: Calculation of the values $z^{(k)}$ and $\sigma^{(k)}$ on the available set $Y^{(k)}$ using formulas (3) and (9).

Step 4: Checking the fulfillment of the inequality $\sigma^{(k)} \leq \sigma_{\max}$. If it is satisfied, the set $Y^{(k)}$ is accepted as the solution, and the search process comes to an end. Otherwise, there is a transition to Step 5.

Step 5: Definition of $Level^{(k+1)}$ for outlier detection:

$$Level^{(k+1)} = 3 \cdot \sigma^{(k)};$$

In order to prevent an infinite loop of iterations, a required verification is carried out:

$$Level^{(k+1)} < Level^{(k)}.$$

If this inequality is not satisfied, then the following is assumed:

$$Level^{(k+1)} = Level^{(k)}/2;$$

Step 6: Definition of a new set $Y^{(k+1)}$ to include those and only those y_j for which

$$\left| y_j - z^{(k)} \right| \leq Level^{(k+1)}.$$

Step 7: Increasing k by 1: k++. Transition to Step 2.

Note that the optimal solution cannot be found in such a manner, as confirmed by numerical calculations (see Section 4).

3. Algorithm for solving the problem

Let us formulate a statement that is the key to the creation of an effective search algorithm for the optimal solution (Eqs. (3)–(7)).

Assertion 1. *Let the set* $Y_{opt} = \left\{ y_{j_1}, \dots, y_{j_{L_{\max}}} \right\}$ *be optimal for a specified sequence of values* $\left\{ y_j \right\}$ *and*

$$y_{\min} = \min\left\{ y_{j_1}, \dots, y_{j_{L_{\max}}} \right\}, y_{\max} = \max\left\{ y_{j_1}, \dots, y_{j_{L_{\max}}} \right\},$$

then the interval $(y_{\min}, y_{\max})$ *does not contain values* y_j *that are not in the set* Y_{opt}.

Proof. In fact, let us assume the opposite: Let $y_j \notin Y_{opt}$, $y_{\min} < y_j < y_{\max}$ and y_k and y_l be values from the set Y_{opt} for which $y_k = y_{\min}$ and $y_l = y_{\max}$. One of these cases is possible:

a. $z < y_j$ and, therefore $0 < \left(y_j - z\right) < (y_l - z) \Rightarrow \left(y_j - z\right)^2 < (y_l - z)^2$,

b. $z \geq y_j$ and, therefore $0 \leq \left(z - y_j\right) < (z - y_k) \Rightarrow \left(y_j - z\right)^2 < (y_k - z)^2$.

In the first case, Case (a), we replace the value y_l in the set Y_{opt} with y_j. In the second case, Case (b), we replace y_k with y_j. In any of the cases, we will have another set of the same length L_{max} for which conditions expressed in Eqs. (3) and (4) are satisfied, but the SD, as follows from Eq. (3), is less than σ_{opt}. Consequently, the set Y_{opt} is not optimal since requirement expressed in Eq. (7) is not satisfied. The contradiction that is reached proves Assertion 1.

Further, note that if $Y_{opt} = \left\{ y_{j_1}, \dots, y_{j_{L_{max}}} \right\}$ is the optimal solution of the problem given by Eqs. (3)–(7), then an arbitrary permutation of the numbers $y_{j_1}, \dots, y_{j_{L_{max}}}$ will also be the optimal solution of the problem described by Eqs. (3)–(7). Thus, the optimal solution does not depend on the arrangement of given numbers y_j. This allows us to arrange the numbers y_j in the order most suitable for searching the optimal solution. Taking advantage of this important circumstance, we arrange the values $\left\{ y_j \right\}$ in the ascending order and we will look for the optimal solution in the ordered sequence. For brevity, the ordered sequence will also be denoted by $\left\{ y_j \right\}$. Thus, $y_1 \le y_2 \le \dots \le y_N$. Moreover, for simplicity of logic, we will assume that all y_j are different, that is,

$$y_1 < y_2 < \dots < y_N \tag{10}$$

Note that due to the formulated Assertion 1, if Y_{opt} is the optimal set and y_{j_1} and $y_{j_{L_{max}}}$ are its smallest and greatest values, respectively, then all values of y_j from the interval $\left(y_{j_1}, y_{j_{L_{max}}} \right)$ belong to Y_{opt}. Consequently, considering Eq. (10), we have $y_{j_{L_{max}}} = y_{j_1 + L_{max} - 1}$ and

$$Y_{opt} = \left\{ y_{j_1}, y_{j_1+1}, \dots, y_{j_1 + L_{max} - 1} \right\}$$

Thus, the optimal solution should be sought in the ascending sequence y_j among all possible sets $\{y_k, \dots, y_{k+L-1}\}$ of length L with k and L satisfying the following conditions:

$$\text{MINOBS} \le L \le N, \tag{11}$$

$$1 \le k \le N - L + 1. \tag{12}$$

Hence, instead of searching for all possible sets of various length, numbering 2^N, for the solution of the problem described by Eqs. (3)–(7), it is sufficient to vary just the two parameters, *k* and *L*, associated with the conditions expressed in Eqs. (11) and (12). The number of pairs of integer numbers *k* and *L* subject to condition in Eqs. (11) and (12) is equal to:

$$(N - \text{MINOBS})(N - \text{MINOBS} + 1)/2.$$

Let $L = l - k + 1$ be the length of an arbitrary set $\left\{ y_k, \dots, y_l \right\}$. We introduce the designations:

$$z(k; L) = \frac{1}{L} \sum_{j=k}^{k+L-1} y_j, \tag{13}$$

$$\sigma^2(k; L) = \frac{1}{L-1} \sum_{j=k}^{k+L-1} \left(y_j - z(k; L)\right)^2. \tag{14}$$

We rewrite conditions given by Eqs. (3) and (4) in the new designations:

$$\sigma^2(k; L) \le \sigma^2_{max}, \tag{15}$$

$$\begin{cases} y_{k+L-1} - z(k; L) \le 3 \cdot \sigma_{max} \\ z(k; L) - y_k \le 3 \cdot \sigma_{max} \end{cases}. \tag{16}$$

Note that the two last inequalities directly follow from Eq. (4), monotony of y_j, and obvious inequalities: $y_k \le z(k; L) \le y_{k+L-1}$.

Remark. *In the conditions expressed in Eqs. (15) and (16), L means the length of the set under checking, and k is the index of the smallest number included in the set. Although "k" and "L" are also encountered as indexes in the sets we use below for monotonically increasing sequences, we hope nevertheless that this will not lead to confusion.*

The following recursive relationships are available, making it possible to find $z(k; L)$ and $\sigma^2(k; L)$ through the calculated values $z(k; L+1)$ и $\sigma^2(k; L+1)$ for seven arithmetic operations:

$$z(k; L) = z(k; L+1) + A_{L+1} \cdot \left(z(k; L+1) - y_{k+L}\right), \tag{17}$$

$$\sigma^2(k; L) = B_{L+1} \cdot \sigma^2(k; L+1) - C_{L+1} \cdot \left(y_{k+L} - z(k; L+1)\right)^2, \tag{18}$$

$$A_L = \frac{1}{L-1}; B_L = \frac{L-1}{L-2}; C_L = \frac{L}{(L-1)(L-2)}. \tag{19}$$

The values of the fractions may be computed in advance as elements of a one-dimensional array. Analogously, the following formulas can make it possible to express $z(k+1; L)$ and $\sigma^2(k+1; L)$ through $z(k; L)$ and $\sigma^2(k; L)$:

$$z(k+1; L) = z(k; L) + A_{L+1} \cdot \left(y_{k+L} - y_k\right), \tag{20}$$

$$\sigma^2(k+1; L) = \sigma^2(k; L) + A_{L+1} \cdot \left(y_{k+L} - y_k\right)\left(y_{k+L} - z(k; L) + D_{L+1}\left(y_k - z(k; L)\right)\right), \tag{21}$$

$$D_L = \frac{L}{L-2}. \tag{22}$$

The algorithm described below is based on the search for all possible pairs (k, L), where L denotes the length of the set to be checked and k is the index of the smallest of the values included in the set. At that, k and L must satisfy conditions Eqs. (11) and (12). The set $\{y_k, \dots, y_{k+L-1}\}$ corresponding to each such pair must be checked for fulfillment conditions (15) and (16).

We organize this search according to the algorithm described below, at each step of which we check the validation of Eqs. (15) and (16) for all possible sets of a certain length. We start the examine process with Step 1, where we check the set of maximum length N. Further, with each next step, we will reduce the length of the sets to be checked by 1.

Step 1: We consider the set of length N. There is only one such set: $\{y_1, \dots, y_N\}$. We check it for fulfillment conditions expressed in Eqs. (15) and (16). If they are satisfied, this set is selected as a solution, and further search stops. Otherwise, transit to Step 2.

Step 2: We consider the sets of length $N-1$. There are two sets of length $N-1$.

$$\{y_1, \dots, y_{N-1}\} \text{ and } \{y_2, \dots, y_N\}.$$

We test each of these sets for compliance with the conditions specified by Eqs. (15) and (16). If none of them satisfies conditions (15) and (16), then we transit to the next step. Otherwise, the following options are available:

- Option 1: if only one set from them is found that satisfies conditions (15) and (16), then it will also be the solution of the stated problem; the search process stops here, where $L_{max} = N - 1$.
- Option 2: if both sets simultaneously satisfy conditions (15) and (16), we will select the set corresponding to the smallest of two values $\sigma^2(1, N-1)$ or $\sigma^2(2, N-1)$, and the search process stops here, where $L_{max} = N - 1$.

Step $N - L + 1$: We consider the sets of length L. If $L < MINOBS$, then the search process stops, and a solution is not found. For $L \geq MINOBS$, we examine $N - L + 1$ sets of length L:

$$\{y_1, \dots, y_L\}, \{y_2, \dots, y_{L+1}\}, \dots, \{y_{N-L+1}, \dots, y_N\}. \quad (23)$$

We check each of these sets for fulfillment of conditions (15) and (16). If any of them does not satisfy these conditions, then we transit to the next step where we consider the sets of length $(L - 1)$. Otherwise, two options are possible:

- Option 1: if only one set from (23) is found that satisfies the conditions of (15) and (16), then it will also be the solution of the stated problem with $L_{max} = L$, and the search process stops here.
- Option 2: if several sets simultaneously satisfy conditions (15) and (16), we chose the set for which the value $\sigma^2(k, L)$ appears smallest as the solution, and the search process stops here, where $L_{max} = L$.

In order to calculate the values $z(k, L)$ and $\sigma^2(k, L)$, we use the recursive formulas (17)–(22) in accordance with a search scheme shown in **Figure 1** where only the $z(k, L)$ is involved.

In accordance with the proposed arrangement, we calculate the values $z(1; N)$ and $\sigma^2(1; N)$ in the first step of the algorithm using formulas expressed in Eqs. (13) and (14), and 4 *N* arithmetic operations are required for this arrangement. In order to find $z(k, L)$ and $\sigma^2(k, L)$ on all subsequent steps, we apply the recursive formulas (17)–(22), making it possible to calculate the values of the specified pairs of

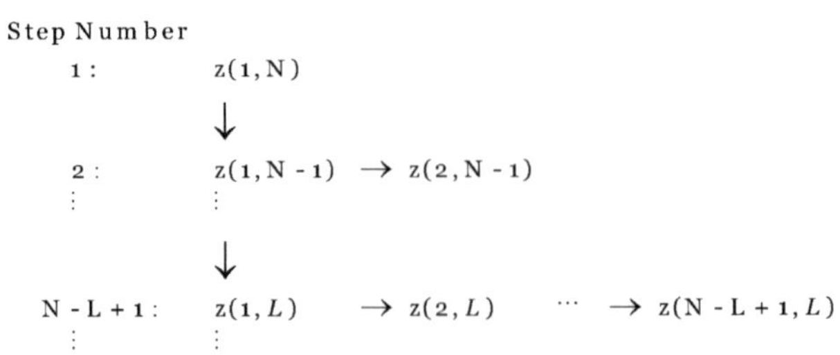

Figure 1.
Scheme of calculations when finding the optimal solution.

variables, each for the seven arithmetic operations, based on the results of the calculations of the preceding step. So, for example, on the second step, proceeding from the known values z(1; N) and $\sigma^2(1; N)$, we find the values z(1; N − 1) and $\sigma^2(1; N-1)$ (vertical arrow on the diagram) by using formulas in Eqs. (17)–(19) and the values $\sigma^2(2; N)$ and $\sigma^2(2; N)$ (horizontal arrow on the diagram) by using Eqs. (20)–(22). In the general case, the transition to the following step in the direction of the vertical arrows (see diagram) is carried out according to formulas (17)–(19), and in the direction of horizontal arrows, according to formulas (20)–(22). The number of arithmetic operations required to find the solution should not exceed:

$$4N + 9((N - L_{max} + 2)(N - L_{max} + 1)/2 - 1) \qquad (24)$$

In the above number of computations, the computational costs of verifying the satisfaction of inequalities (15) and (16) are also considered, which comprise from 0 to 2 arithmetic operations.

4. Results of calculations using algorithm presented in Section 3

We test the algorithms discussed above on the real data obtained by the ONSA station that is a part of the IGS network [2]. These data are included in the distribution kit of the installation software package [3] and available for usage. We consider measurement data received from global positioning system (GPS) satellite with system number PRN = 12 for 2010, day 207 to check the efficiency of the proposed algorithm described in Section 3. **Figure 2** plots the values of the Melbourne-Wübbena combination over a time interval of 89.5 min (N = 180). The index numbers j of time epochs counting from the beginning of a 24-h period with a 30-second interval are plotted on the horizontal axis. The values y_j of the combination are plotted on the vertical axis and are expressed in cycles with wavelength $\lambda_5 \approx 0.86$ [3]. **Figure 3** shows the values of deviations from the mean of the data cleared of outliers using the algorithms described in Sections 2 and 3. In both cases, σ_{max} = 0.6 and MINOBS = 10. The values $y_j - z$ are plotted on the vertical axes in cycles with wavelength λ_5 and the index numbers j of epochs on the horizontal axis.

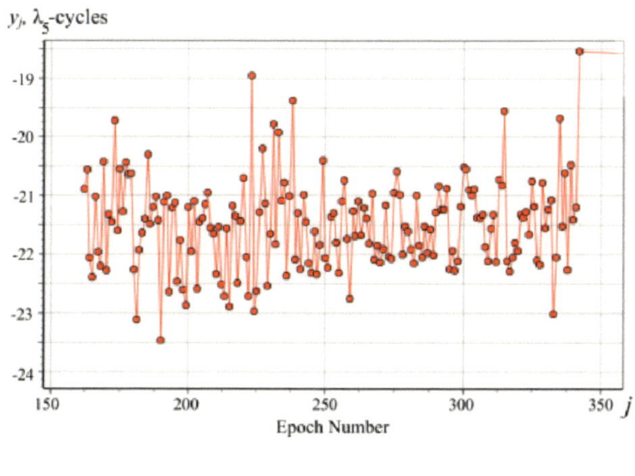

Figure 2.
Melbourne-Wübbena combination for ONSA station (GPS satellite, PRN = 12 for 2010, day 207).

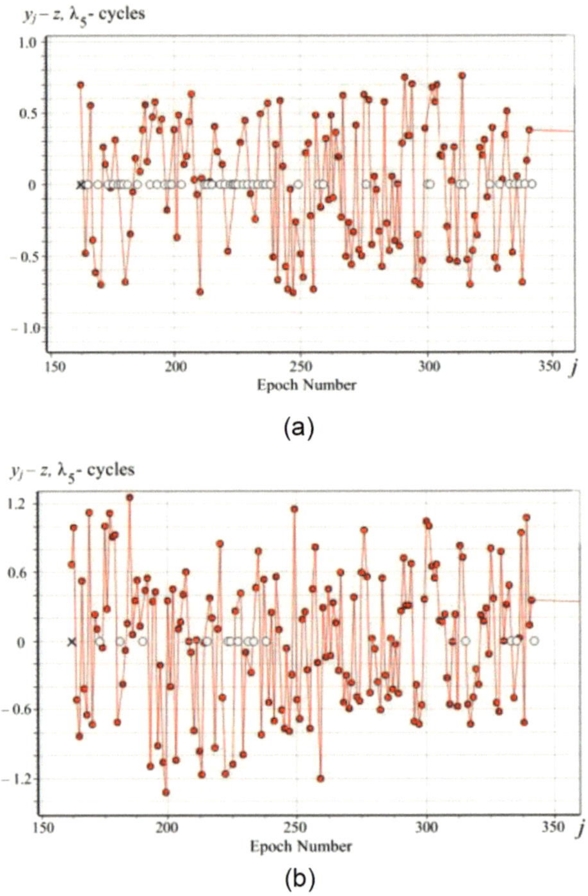

Figure 3.
(a) Deviations of values of the Melbourne-Wübbena combination from the mean value after data cleaning from outliers using the algorithm described in Section 2. (b) Deviations of values of the Melbourne-Wübbena combination from the mean value after data cleaning from outliers using the developed algorithm (see Section 3).

Epochs in which the data were rejected are designated by white circles. In the first case (see **Figure 3a**), 47 data of the measurements were rejected, which are 26.1% of the total amount of data. In the second case (see **Figure 3b**), 14 of these measurements were rejected (7.8%), which are almost 18% less than in the previous calculation.

We also provide similar results for data obtained by TLSE station, which is also included in the IGS network. We consider measurement data from GLONASS, Russia satellite with system number PRN = 1 for 2010, day 207. **Figure 4** shows the values of the Melbourne-Wübbena combination over a time interval of 65.5 min (N = 132). **Figure 5** plots the values of deviations from the mean value of the data cleared of outliers using the algorithms described in Sections 2 and 3, respectively. Parameters σ_{max} and MINOBS are the same as in the previous calculation example. In the first case (see **Figure 5a**), 41 data of the measurements were discarded, which are 31% of the total amount data. In the second case (see **Figure 5b**), 8 of these measurements were rejected (6%), which are 25% less than in the previous calculation.

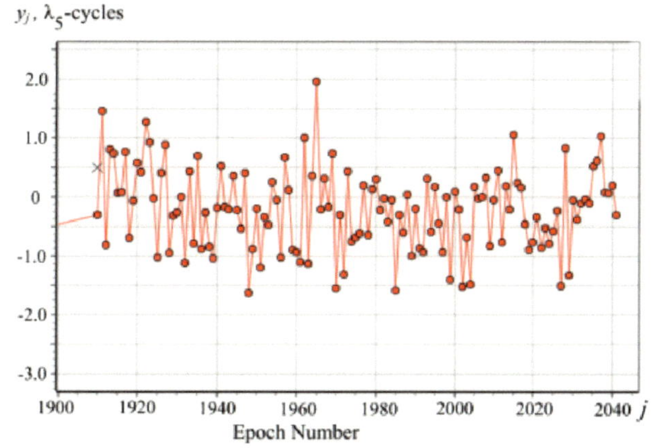

Figure 4.
Melbourne-Wübbena combination for TLSE station (GLONASS satellite, PRN = 1 for 2010, day 207).

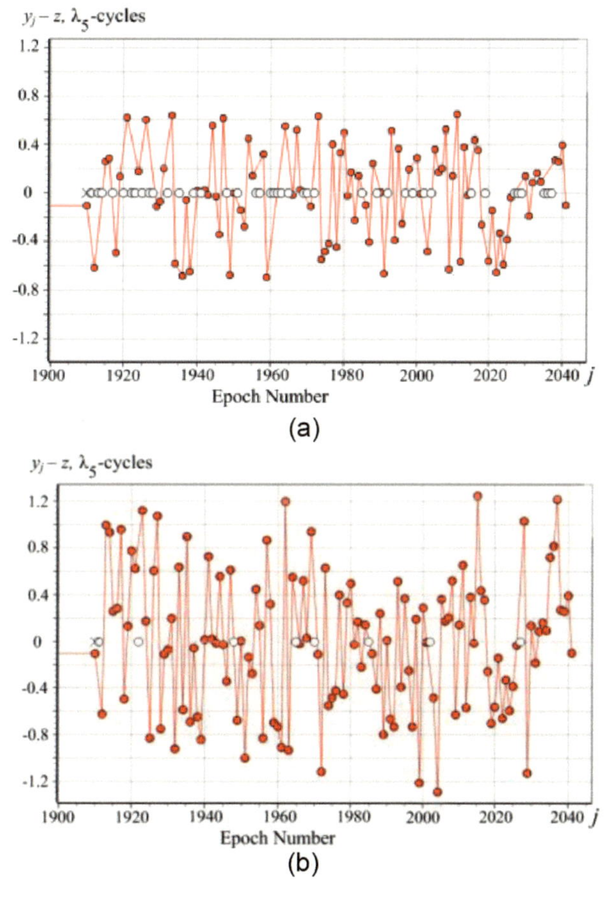

Figure 5.
(a) Deviations of values of the Melbourne-Wübbena combination from the mean value after data cleaning from outliers using the algorithm described in Section 2. (b) Deviations of values of the Melbourne-Wübbena combination from the mean value after data cleaning from outliers using the developed algorithm (see Section 3).

5. Mathematical prerequisites for modifying of existing algorithm

Note that the number of arithmetic operations required to find the optimal solution according to the algorithm described in Section 3 depends on the L_{max}, which is the length of the solution. As can be seen from Eq. (24), the smaller the length of the found solution (i.e., the larger the number of detected outliers), the more arithmetic operations are required to find it. This number of arithmetic operations is estimated to be of order $N + (N - L_{max})^2$. Note that the expression in the parentheses herein is equal to the number of outliers detected. Thus, if the number of outliers in the original data series is comparable to N, it will take $\sim N^2$ arithmetic operations to find the optimal solution. In particular, to make certain that there is no solution (e.g., in the case where the data contain a trend), $\sim N^2$ arithmetic operations will also be required. In Section 6, we will modify the existing algorithm and describe fast outlier search algorithm that requires $\sim N\log_2 N$ arithmetic operations.

The necessary preparations are given in this section. Note that in this and the next sections we are dealing with the sequence $\left\{y_j\right\}_{j=1}^{N}$ arranged in the ascending order.

Assertion 2. *Let* $\left\{y_j\right\}_{j=1}^{N}$ *be monotonically increasing sequence. The following inequality is true:*

$$\sigma^2(k; L+1) \geq \min\left\{\sigma^2(k; L),\ \ \sigma^2(k+1; L)\right\}. \tag{25}$$

Proof. From the monotonicity of the sequence y_j and the definition $z(k, L+1)$ [see Eq. (13)],

$$y_k \leq z(k, L+1) \leq y_{k+L}. \tag{26}$$

One of two cases is possible:

a. $2z(k, L+1) \leq y_{k+L} + y_k, \Rightarrow z(k, L+1) - y_k \leq y_{k+L} - z(k, L+1),$

b. $2z(k, L+1) > y_{k+L} + y_k, \Rightarrow z(k, L+1) - y_k > y_{k+L} - z(k, L+1).$

Suppose, for example, the case (a) holds. Let us show that in this case

$$\sigma^2(k; L+1) \geq \sigma^2(k; L). \tag{27}$$

At first, we will show that:

$$|y_j - z(k, L+1)| \leq y_{k+L} - z(k, L+1); j = k, \ldots, k+L. \tag{28}$$

Truly, inequalities:

$$y_k \leq y_j \leq y_{k+L}; j = \ldots, k+L,$$

and the above inequality derived in Case (a) implies:

$$z(k, L+1) - y_j \leq z(k, L+1) - y_k \leq y_{k+L} - z(k, L+1), \tag{29}$$

$$y_j - z(k, L+1) \leq y_{k+L} - z(k, L+1). \tag{30}$$

These inequalities, in turn, imply Eq. (28). Next let us prove (27). This inequality is expanded as follows:

$$\frac{1}{L}\sum_{j=k}^{k+L}\left(y_j - z(k; L+1)\right)^2 \geq \frac{1}{L-1}\sum_{j=k}^{k+L-1}\left(y_j - z(k; L)\right)^2.$$

Substituting here of the expression (17) in place of $z(k; L)$ and writing for brevity, z instead of $z(k; L+1)$, we get inequality:

$$(L-1)\sum_{j=k}^{k+L}\left(y_j - z\right)^2 \geq L\sum_{j=k}^{k+L-1}\left(y_j - z - \frac{z - y_{k+L}}{L}\right)^2. \tag{31}$$

Transform the right-hand side of this inequality:

$$\begin{aligned}\mathrm{RHS}(31) &= L\sum_{j=k}^{k+L}\left(y_j - z + \frac{y_{k+L} - z}{L}\right)^2 - \frac{(L+1)^2}{L}\left(y_{k+L} - z\right)^2 \\ &= L\sum_{j=k}^{k+L}\left(y_j - z\right)^2 + \frac{(L+1)}{L}\left(y_{k+L} - z\right)^2 - \frac{(L+1)^2}{L}\left(y_{k+L} - z\right)^2.\end{aligned}$$

Here we take into account the equality: $\sum_{j=k}^{k+L}\left(y_j - z\right) = 0$. Next, we have:

$$\mathrm{RHS}(31) = L\sum_{j=k}^{k+L}\left(y_j - z\right)^2 - (L+1)\left(y_{k+L} - z\right)^2.$$

Substituting this expression in Eq. (31), we get inequality

$$\sum_{j=k}^{k+L}\left(y_j - z\right)^2 \leq (L+1)\left(y_{k+L} - z\right)^2,$$

that is true due to Eq. (28). Thus, Eq. (27) is proved for case (a). Analogously, case (b) is considered.

We introduce the notation:

$$\sigma^2_{min}(L) = \min_{1 \leq k \leq N-L+1} \sigma^2(k, L). \tag{32}$$

Assertion 3. *The following inequalities hold:*

$$\sigma^2_{min}(N) \geq \sigma^2_{min}(N-1) \geq \ldots \geq \sigma^2_{min}(MINOBS). \tag{33}$$

That is, the sequence $\sigma^2_{min}(L)$ *monotonically decreases when L decreases from N to MINOBS.*

Proof. Assertion 2 and definition of $\sigma^2_{min}(L)$ expressed in Eq. (32) imply:

$$\sigma^2(k, L+1) \geq \min\left\{\sigma^2(k, L), \sigma^2(k+1, L)\right\} \geq \sigma^2_{min}(L).$$

Since k is chosen arbitrarily, then for all L = MINOBS, ..., N − 1 the following inequalities hold:

$$\sigma^2_{min}(L+1) \geq \sigma^2_{min}(L),$$

which proves Assertion 3.

Assertion 3 implies the following corollary.

Corollary 1. *If the inequality*

$$\sigma^2_{min}(L_0) > \sigma^2_{max}. \quad (34)$$

holds for some L_0*, then for existence of the solution* $Y_L = \{y_k, y_{k+1}, \dots, y_{k+L-1}\}$ *for the problem expressed in Eqs. (3)–(7), it is necessary that the length L of the set* Y_L *satisfies the condition* $L < L_0$.

Proof. Let us assume that $L \geq L_0$. Assertion 3 on account of monotony of $\sigma^2_{min}(\cdot)$, expressed in Eq. (33) implies the following inequalities $\sigma^2_{min}(L) \geq \sigma^2_{min}(L_0) > \sigma^2_{max}$ for all $L \geq L_0$. From this, it follows that for any set $Y_L = \{y_k, y_{k+1}, \dots, y_{k+L-1}\}$ we will have $\sigma^2(k, L) \geq \sigma^2_{min}(L) > \sigma^2_{max}$. Thus, any of sets Y_L of length $L \geq L_0$ does not satisfy the condition in Eq. (15) and, therefore, cannot be a solution of the problem, expressed in Eqs. (3)–(7).

In particular, we have come to the next important result. If, for example, the inequalities $\sigma^2_{min}(MINOBS) > \sigma^2_{max}$ are fulfilled, then the solution for the problem described in Eqs. (3)–(7) does not exist.

In the above-described procedure for solving problem (3)–(7), it takes $\sim N^2$ arithmetic operations to make certain that the solution not exists. Taking into account Assertion 3 and Corollary 1, the search procedure may begin by checking the conditions.

$$\sigma^2_{min}(N) = \sigma^2(1, N) \leq \sigma^2_{max} \text{ and } \sigma^2_{min}(MINOBS) \leq \sigma^2_{max}.$$

This will require approximately $\sim N$ arithmetic operations. If none of these conditions are fulfilled, the solution search must stop because the solution does not exist. As a result, only $\sim N$ arithmetic operations are required to ensure that there is no solution.

6. Fast outlier search algorithm

The above proposed search procedure consists in the calculating values of $z(k, L)$ and $\sigma^2(k, L)$ using recurrent formulas (17)–(22) and checking at every k and L the fulfillment of the inequalities (11) and (12). The algorithm complexity is estimated by value of $\sim(N + N^2_{Outlier})$, where $N_{Outlier}$ is the number of outliers found. If it is known a priori that there are few outliers in the measurement data, then the search algorithm for the optimal solution that described in Section 3 can be applied. If the measurement data contain an N-comparable number of outliers, the complexity of such an algorithm will be estimated by about N^2. It is namely for such type of data we below describe a modified outlier search algorithm with complexity of about $N\log_2 N$.

First of all, note one property that is the key to the construction of a fast outlier search algorithm. Note that if the inequality (15) holds for some set of length L + 1, then there exists a set of length L for which the inequality (15) is valid too. Truly, let assume for some k the inequality $\sigma^2(k, L+1) \leq \sigma^2_{max}$ holds. This inequality and Eq. (25) imply

$$\min\left\{\sigma^2(k; L),\ \sigma^2(k+1; L)\right\} \le \sigma^2_{max}.$$

From this, it follows that

$$\sigma^2(k; L) \le \sigma^2_{max} \text{ and/or } \sigma^2(k+1; L) \le \sigma^2_{max}. \tag{35}$$

This means that at least one of these sets $\{y_k, \dots, y_{k+L-1}\}$ and $\{y_{k+1}, \dots, y_{k+L}\}$ with length of L satisfies conditions expressed in Eq. (15).

However, this property is not true when checking the conditions expressed in Eq. (16). In other words, if these conditions are fulfilled for any set of length L + 1, it might happen that none of the sets of length L may satisfy them. This fact is a significant obstacle to increasing the rate of outlier detection that is necessary when processing a large amount of data with a large number of rough measurements. To overcome this obstacle, we will make the condition expressed in Eq. (16) weaker.

First of all, note that if for some set $\{y_k, \dots, y_{k+L-1}\}$, the both conditions expressed in Eq. (16) are fulfilled, then the following inequality will hold:

$$y_{k+L-1} - y_k \le 6\sigma_{max}. \tag{36}$$

Consider a problem with condition expressed in Eq. (36) instead of conditions expressed in Eq. (16).

Remark. *Recall that in this condition L means the length of the set under checking, and k is the index of the smallest number included in the set. Although "k" and "L" are also encountered as indexes in the sets we use hereinafter, we hope nevertheless that this will not lead to confusion.*

It is easily seen that condition expressed in Eq. (36) for an arbitrary set $\{y_k, \dots, y_{k+L}\}$ of length L + 1 implies the fulfillment this condition for each of the sets $\{y_k, \dots, y_{k+L-1}\}$ and $\{y_{k+1}, \dots, y_{k+L}\}$ of length L. In fact, the fulfillment condition (36) for the set $\{y_k, \dots, y_{k+L}\}$ means the fulfillment of inequality $y_{k+L} - y_k \le 6\sigma_{max}$ from which due to monotony of y_j, the inequalities imply $y_{k+L} - y_{k+1} \le 6\sigma_{max}$ and $y_{k+L-1} - y_k \le 6\sigma_{max}$.

Thus, we have established the validity of the following assertion.

Assertion 4. *If the set $\{y_k, \dots, y_{k+L}\}$ of length L + 1 satisfies conditions (15) and (36), then at least one of the two sets $\{y_k, \dots, y_{k+L-1}\}$ or $\{y_{k+1}, \dots, y_{k+L}\}$ of length L also satisfies conditions (15) and (36).*

Based on this statement, we can formulate the following:

Assertion 5. *Solution for the problem (15) + (36) can be found for $\sim N \log_2 N$ arithmetic operations.*

Proof. Let us consider the sequence of steps.

Step 0: Consider the segment $\left[N^{(0)}_{Left}, N^{(0)}_{Right}\right]$ of numerical axis, where $N^{(0)}_{Left}$ = MINOBS, $N^{(0)}_{Right}$ = N. In this step, there is one set $\{y_1, \dots, y_N\}$ of length N. We check it for fulfillment of the conditions expressed in Eqs. (15) and (36) with k = 1, L = N. If they are satisfied, this set is the solution, and our search stops. Otherwise, we pass to considering of (N − MINOBS − 1) sets of length MINOBS. We check each of these sets for fulfillment of conditions expressed in Eqs. (15) and (36). If none of them satisfy these conditions, then we stop the search process and conclude that the solution does not exist. Otherwise, once we find the set of length MINOBS satisfying conditions (15) and (36), we transit to Step 1.

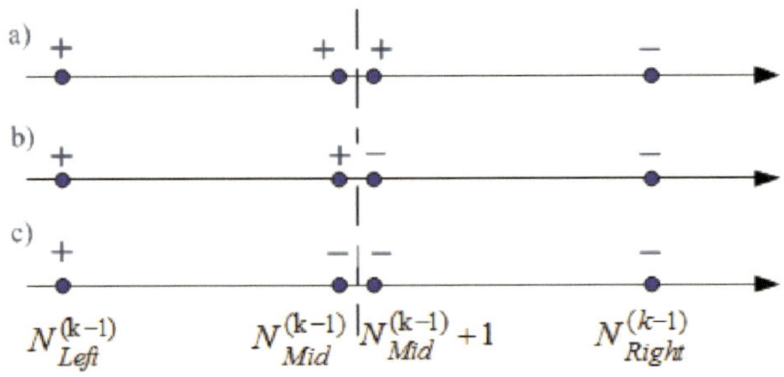

Figure 6.
Three possible cases for the proposed search. In case (a) we go to the right-hand side range (range for length of sets) to find a solution, in case (c) we go to the left-hand side range to find a solution, in case (b) we look for a solution with length $L = N_{Mid}^{(k-1)}$ *and the search ends.*

Step 1: Step 1 is the same as Step k described below for k = 1

...

Step k: On the *k*th step, where (k ≥ 1), we consider a segment $\left[N_{Left}^{(k-1)}, N_{Right}^{(k-1)}\right]$, which we halve, as a result we receive two segments $\left[N_{Left}^{(k-1)}, N_{Mid}^{(k-1)}\right]$ and $\left[N_{Mid}^{(k-1)}+1, N_{Right}^{(k-1)}\right]$, where $N_{Mid}^{(k-1)} = N_{Left}^{(k-1)} + \left[\left(N_{Right}^{(k-1)} - N_{Left}^{(k-1)}\right)/2\right]$, $[\cdot]$ designate integral part of a number. Next, we check the sets of length $N_{Mid}^{(k-1)}$ and $N_{Mid}^{(k-1)} + 1$ for fulfillment of conditions (15) and (36). The following three cases are possible, schematically shown in **Figure 6**. The sign "−" above the point means that for none of the sets of the corresponding length the conditions (15) + (36) are satisfied, the sign "+" vice versa, that is, there is at least one set of the corresponding length satisfying (15) + (36). In the case of (a), we set $N_{Left}^{(k)} = N_{Mid}^{(k-1)} + 1$, $N_{Right}^{(k)} = N_{Right}^{(k-1)}$ and transit to the (k + 1)-th step; in the case of (c), we set $N_{Left}^{(k)} = N_{Left}^{(k-1)}$, $N_{Right}^{(k)} = N_{Mid}^{(k-1)}$ and transit to the (k + 1)-th step; in the case of (b), we search the solution (15) + (36) with minimal value of $\sigma^2(k, L)$ with $L = N_{Mid}^{(k-1)}$; the algorithm is terminated.

The search process will continue until either case (b) or until the length of the segment $\left[N_{Left}^{(k)}, N_{Right}^{(k)}\right]$ is less than or equal to 1. In either case, the total number of steps will not exceed the number of $\log_2(N - \text{MINOBS})$. Since ~N operations are performed in each step, the search process is guaranteed to be terminated in $\sim N\log_2 N$ arithmetic operations.

7. Search an unknown trend in the power polynomial class

The need to process GNSS measurements including a trend on which noise and outliers are superimposed arises at different processing stages of the application process. As already stated above, satellite clock corrections contain a linear trend. In some cases, it may not be known, and then, one has to search for it, for example, by the least square method. The presence of outliers in the measurement data is a significant obstacle to accurate determination of drift and offset parameters of satellite clocks. Other examples are linear combinations of code and phase data on two carriers [3]. To obtain high accuracy results, it is necessary to detect outliers against an unknown trend and remove them from further processing. This is the subject of this section.

7.1 Statement of the problem

Consider the problem of outlier detecting in data presented in the form of Eq. (1), recall that:

$$y_j = f_j + z + \xi_j; j = 1 \ldots N.$$

The procedure described above for finding the optimal solution in an ordered series of numbers may not produce an adequate result if applied to data containing an unknown trend. For example, there may be no solution, and all data will be defined as outliers. In order to detect outliers in a series of numbers with a trend using the algorithm described above, it is necessary to find a suitable approximation of an unknown function f_j and subtract it from the data set. Searching for this approximation is usually done by selecting functions from some functional class. The choice of the functional class depends on the task. In some cases, these may be power polynomial, in other cases, trigonometric polynomial, etc. The presence of outliers in the data measurements makes it much more difficult to find such an approximation. In this section, we will describe the general approach to solving the problem and look for suitable approximations of trend f_j in the class of power polynomial with real coefficients:

$$P_{n,j}\left(\vec{a}\right) = a_n(j/N)^n + a_{n-1}(j/N)^{n-1} + \ldots + a_0, \tag{37}$$

where n is the polynomial degree, and $\vec{a} = \{a_0, \ldots, a_n\}$ is vector of coefficients.

Thus, the problem consists in the creation of an algorithm for searching the trend in the class of power polynomial and detecting outliers in specified data series y_j represented in Eq. (1).

7.2 Minimizing set of specified length L

Before we turn to the trend search algorithm construction, we will define the so-called minimizing set of given length L, which plays an essential role in the trend search. In addition, in Section 7.3, we will describe a search algorithm for such set based on the recurrent formulas (17)–(19) and (20)–(22).

Let $Y_L = \left\{y_{j_1}, \ldots, y_{j_L}\right\}$ be an arbitrary set of length L, composed of the values of a numerical series $\left\{y_j\right\}_{j=1}^{N}$, the monotony is not supposed. The mean and the SD values for it are denoted by z_{Y_L} and σ_{Y_L}. These values are calculated by standard formulas:

$$z_{Y_L} = L^{-1} \sum_{j \in \{j_1, \ldots, j_L\}} y_j, \tag{38}$$

$$\sigma^2_{Y_L} = (L-1)^{-1} \sum_{j \in \{j_1, \ldots, j_L\}} \left(y_j - z_{Y_L}\right)^2. \tag{39}$$

Definition 2. *Given L for a specified sequence of values* $\left\{y_j\right\}_{j=1}^{N}$ *the set of values:*

$$Y_{L,\min} = \left\{y_{j_1}, \ldots, y_{j_L}\right\},$$

at which the minimum value of $\sigma^2_{Y_L}$ *defined in Eq. (39) is reached will be called the minimizing set of length L. The corresponding mean and SD values are denoted by* $z_{Y_{L,\min}}$ *and* $\sigma_{Y_{L,\min}}$.

According to this definition, we have

$$z_{Y_{L,\min}} = L^{-1} \sum_{j \in \{j_1, \dots, j_L\}} y_j, \tag{40}$$

$$\sigma^2_{Y_{L,\min}} = (L-1)^{-1} \sum_{j \in \{j_1, \dots, j_L\}} \left(y_j - z_{Y_{L,\min}}\right)^2 = \min_{Y_L} \left\{\sigma^2_{Y_L}\right\}. \tag{41}$$

Minimum in Eq. (41) is searched by all kinds of sets of length L composed of numbers of series $\left\{y_j\right\}_{j=1}^{N}$.

Note that the numbers y_j are not supposed to be in the ascending order.

Next, we will formulate and prove a statement similar to Assertion 1, which will allow us, when searching for a minimizing set, to proceed from the original series to its ordered permutation.

Assertion 6. Let $Y_{L,\min} = \left\{y_{j_1}, \dots, y_{j_L}\right\}$ be a minimizing set of length L *for a given sequence of values* $\left\{y_j\right\}_{j=1}^{N}$ and

$$y_{\min} = \min\left\{y_{j_1}, \dots, y_{j_L}\right\}, y_{\max} = \max\left\{y_{j_1}, \dots, y_{j_L}\right\},$$

then the interval $(y_{\min}, y_{\max})$ *does not contain values* y_j *that are not in the set* $Y_{L,\min}$.

Proof. Let us assume the opposite: Let $y_j \notin Y_{L,\min}$, $y_{\min} < y_j < y_{\max}$. Let $y_k, \dots, y_{k+L-1}$ is a permutation of the numbers $y_{j_1}, \dots, y_{j_L}$ in the ascending order; then $y_k = y_{\min}$ and $y_{k+L-1} = y_{\max}$. One of these cases is possible:

a. $y_j < z_{Y_{L,\min}}$ and therefore subject to inequality $y_k < y_j$, we have

$$\left(z_{Y_{L,\min}} - y_k\right) > \left(z_{Y_{L,\min}} - y_j\right) \Rightarrow \left(z_{Y_{L,\min}} - y_k\right)^2 > \left(z_{Y_{L,\min}} - y_j\right)^2$$
$$\Rightarrow \left(z_{Y_{L,\min}} - y_j\right)^2 - \left(z_{Y_{L,\min}} - y_k\right)^2 < 0 \tag{42}$$

b. $y_j \geq z_{Y_{L,\min}}$ and therefore subject to inequality $y_{k+L-1} > y_j$, we have

$$\left(y_{k+L-1} - z_{Y_{L,\min}}\right) > \left(y_j - z_{Y_{L,\min}}\right) \Rightarrow \left(y_j - z_{Y_{L,\min}}\right)^2 - \left(y_{k+L-1} - z_{Y_{L,\min}}\right)^2 < 0 \tag{43}$$

In the first case, Case (a), we replace the value y_k in the set $\left\{y_k, \dots, y_{k+L-1}\right\}$ with y_j. In the second case, Case (b), we replace the value y_{k+L-1} with y_j. We want to show that doing such replacement, the value $\sigma^2_{Y_{L,\min}}$ expressed in Eqs. (40) and (41) will decrease. This will mean that $Y_{L,\min}$ is not a minimizing set.

Suppose Case (a). For brevity, we will write below z instead of $z_{Y_{L,\min}}$ and σ^2 instead of $\sigma^2_{Y_{L,\min}}$. Denote $\tilde{z}$ and $\tilde{\sigma}^2$, the similar values obtained after replacement y_k with y_j. We have:

$$z = L^{-1}\left(y_k + \dots + y_{k+L-1}\right), \tag{44}$$

$$\sigma^2 = (L-1)^{-1}\left((y_k - z)^2 + \ldots + (y_{k+L-1} - z)^2\right), \tag{45}$$

and

$$\tilde{z} = L^{-1}\left(y_{k+1} + \ldots + y_{k+L-1} + y_j\right), \tag{46}$$

$$\tilde{\sigma}^2 = (L-1)^{-1}\left((y_{k+1} - \tilde{z})^2 + \ldots + (y_{k+L-1} - \tilde{z})^2 + \left(y_j - \tilde{z}\right)^2\right). \tag{47}$$

We want to show that

$$\tilde{\sigma}^2 < \sigma^2. \tag{48}$$

Eqs. (44) and (46) imply:

$$\tilde{z} = z + L^{-1}\left(y_j - y_k\right). \tag{49}$$

Modify $\tilde{\sigma}^2$ expressed in Eq. (47) taking account of Eq. (49):

$$\tilde{\sigma}^2 = (L-1)^{-1}\left(\left(y_{k+1} - z - L^{-1}\left(y_j - y_k\right)\right)^2 + \ldots \right.$$
$$\left. + \left(y_{k+L-1} - z - L^{-1}\left(y_j - y_k\right)\right)^2 + \left(y_j - z - L^{-1}\left(y_j - y_k\right)\right)^2\right).$$

After simplification with taking into account Eq. (45), we get from here:

$$\tilde{\sigma}^2 = \sigma^2 + (L-1)^{-1}\left[\left(y_j - z\right)^2 - (y_k - z)^2 - L^{-1}\left(y_j - y_k\right)^2\right].$$

From (42), it follows (recall that we write z instead of $z_{Y_{L,\min}}$) that the expression in the square brackets is strictly less than zero. Thus, inequality (48) is proven. From this follows that the set $Y_{L,\min}$ is not minimizing one because the condition (41) is not met. Thus, we have arrived at a contradiction that proves the validity of the formulated Assertion 6. Case (b) is considered similarly.

7.3 $Y_{L,\min}$ search algorithm

Assertion 6 is much like Assertion 1, which made it possible to go from an arbitrary numerical series to an ordered one for the optimal solution search (see Section 3). Similarly, Assertion 6 makes it possible to go to an ordered number series to find the minimizing set of numbers of a given length. In our case, considerations similar to those presented in Section 3 may be made when searching for a minimizing set. Namely, if $Y_{L,\min} = \left\{y_{j_1}, \ldots, y_{j_L}\right\}$ is the minimizing set of length L, then an arbitrary permutation of the numbers $y_{j_1}, \ldots, y_{j_L}$ will also be the minimizing set. Thus, the process for searching of $Y_{L,\min}$ does not depend on the arrangement of given numbers y_j. This allows us to arrange them in the order most suitable for searching the minimizing set. We arrange the values $\left\{y_j\right\}$ in the ascending order, and we will look for the minimizing set in the ordered sequence. As in Section 3, for brevity, the ordered sequence will also be denoted by $\left\{y_j\right\}$. For simplicity of logic, we will assume that all y_j are different, that is, Eq. (10) holds. Rewrite it for convenience:

$$y_1 < y_2 < \dots < y_N$$

Note that due to Assertion 6, if $Y_{L,\min}$ is the minimizing set and y_{j_1} and y_{j_L} are its smallest and greatest values, respectively, then all values of y_j from the interval $\left(y_{j_1}, y_{j_L}\right)$ belong to $Y_{L,\min}$. Consequently, considering Eq. (10), we have $y_{j_L} = y_{j_1+L-1}$ and therefore

$$Y_{L,\min} = \left\{y_{j_1}, y_{j_1+1}, \dots, y_{j_1+L-1}\right\}.$$

Thus, in order to find a minimizing set of length L, it is sufficient for us to check N − L + 1 sets

$$\{y_1, \dots, y_L\}, \{y_2, \dots, y_{L+1}\}, \dots, \{y_{N-L+1}, \dots, y_N\}.$$

and choose from them the one that has minimal SD value. In the minimizing set searching procedure, we use the appropriate designations $z(k, L)$ and $\sigma^2(k, L)$ expressed by Eqs. (13) and (14) to denote mean and square SD in the case of sets composed of the ascending ordered values.

The calculation of $z(k, L)$ and $\sigma^2(k, L)$ when searching $Y_{L,\min}$ can be carried out in the manner similar to that of the optimal solution according to the schematic, in which only the $\sigma^2(k, L)$ is shown:

Length of Set			
N	$\sigma^2(1, N)$		
	$\downarrow$		
N - 1	$\sigma^2(1, N-1)$		
$\vdots$	$\vdots$		
	$\downarrow$		
L	$\sigma^2(1, L)$	$\rightarrow \sigma^2(2, L)$ $\cdots$	$\rightarrow \sigma^2(N-L+1, L)$

Figure 7.
Scheme of calculations when finding the minimizing set of length L.

At first, we carry out the transitions in the direction of the vertical arrows (see diagram in **Figure 7**) and calculate sequential values $\sigma^2(1, N), \sigma^2(1, N-1), \dots, \sigma^2(1, L)$ using formula (17)–(19). Finally, we go in the direction of the horizontal arrows and calculate values of $\sigma^2(1, L), \sigma^2(2, L), \dots, \sigma^2(N-L+1, L)$ using formula (20)–(22) and choose from them the minimal one.

7.4 Trend search algorithm

In order to correctly detect outliers in measurement data that include an unknown trend, it is necessary to find and remove trend from the original data. The problem in determining of unknown trend is to find a suitable profile for the measurement data by adjusting the fitting parameters. This implies, in turn, the needs to select from the specified series y_j the "right" reference values $\left\{y_{j_1}, \dots, y_{j_L}\right\}$ used for fitting. If the data do not contain coarse measurements, all N values of the original data can be treated as reference ones and used for fitting. If the data contain rough measurements, the participation all or some of those values in the fit will result in a fatal distortion of the trend function and, as a result, an incorrect determination of the outliers. Thus, when determining the trend, it is necessary to exclude rough measurements from the number of reference values used for fitting and by which the trend approximations are built. We have the vicious circle. To determine outliers, it is necessary to find a proper trend approximation, and to find an exact trend approximation, we need to find all outliers and remove them from the values used for fitting. Another vicious circle is obtained when trying to choose the number of values for fitting. If we take it too small to minimize the possibility for outliers to be included into the subset of values for fitting, the data fit may not be accurate enough for the rest values of the data outside of the subset. If, on the contrary, we take rather large number of points for fitting, the outliers may be among them, and we will also get a bad approximation of the trend.

We describe herein a strategy for finding an unknown trend and detecting outliers. This strategy assumes that the number of outliers in the series presented by Eq. (1) does not exceed a certain value N_{maxout} known a priori. Thus, we can suppose that the number of "right" values suitable for fitting in the series y_j is not less than $N - N_{maxout}$.

Below L is supposed to be fixed and associated with the number of the reference values of the series (1) used for fitting, and $L \le N - N_{maxout}$.

Let us consider the following algorithm. It contains internal iterations, which we will denote with the upper index "s" in parentheses.

Step 0: n = 0. We set some L, satisfying the condition $L \le N - N_{maxout}$.

Step 1: n++; s = 0; $\text{flag}_j^{(0)} = 1$.

Step 2: s++. We fit polynomial to the data set and find fitting coefficients $\vec{a}^{(s)} = \left\{a_0^{(s)}, \dots, a_n^{(s)}\right\}$:

$$\vec{a}^{(s)} = \underset{a_0, \dots, a_{n-1}, a_n}{\arg\min}\ \Phi^{(s-1)}\left(\vec{a}\right), \tag{50}$$

$$\Phi^{(s-1)}\left(\vec{a}\right) \triangleq \sum_{j=1}^{N}\left(y_j - P_{n,j}\left(\vec{a}\right)\right)^2 \cdot \text{flag}_j^{(s-1)}. \tag{51}$$

Step 3: Consider the values $\hat{y}_j^{(s)}$ obtained after subtraction from y_j the polynomial values with coefficients found in Step 2:

$$\hat{y}_j^{(s)} = y_j - P_{n,j}\left(\vec{a}^{(s)}\right). \tag{52}$$

Using the algorithm described in Section 7.3, we find from the numbers $\left\{\hat{y}_j^{(s)}\right\}_{j=1}^{N}$ defined in Eq. (52) a minimizing set $Y_{L,\min}^{(s)} = \left\{\hat{y}_{j_1^{(s)}}^{(s)}, \dots, \hat{y}_{j_L^{(s)}}^{(s)}\right\}$ of length L. For this set, we calculate the corresponding values of the mean $z_{Y_{L,\min}^{(s)}}$ and SD $\sigma_{Y_{L,\min}^{(s)}}$ according to the formulas (40) and (41) in which $\{j_1, \dots, j_L\}$ is replaced by $\left\{j_1^{(s)}, \dots, j_L^{(s)}\right\}$. We redefine the reference points for the next fit process (Step 2) by marking them with $\text{flag}_j^{(s)}$:

$$\text{flag}_j^{(s)} = \begin{cases} 1, & \text{if } j \in \left\{j_1^{(s)}, \dots, j_L^{(s)}\right\} \\ 0, & \text{otherwise} \end{cases}. \tag{53}$$

We transit to Step 2 and do so until the convergence of the $\sigma_{Y^{(s)}_{L,\min}}$, s = 1,2, …, is achieved. Note that we will consider the issue of convergence further in Section 7.5. After reaching convergence of values $\sigma_{Y^{(s)}_{L,\min}}$, we transit to Step 4 for outlier detection.

Step 4: Searching the optimal solution for data set $\hat{y}_j^{(s)} = y_j - P_{n,j}\left(\vec{a}^{(s)}\right)$, j = 1, …, N.

We find the optimal solution for values $\hat{y}_j^{(s)}$ using the algorithm of Section 3 and determine the number N_{out} of outliers detected. If it turns out that $N_{out} \le N_{maxout}$, then solution is considered found; searching process stops: $f_j = P_{n,j}\left(\vec{a}^{(s)}\right)$. Otherwise, if $N_{out} > N_{maxout}$, then we transit to Step 1.

We do this until we find a solution or until n reaches some preset value N_{max} (e.g., 10). In this case, probably, we may need to select a different functional class to search for a trend.

Example. Let us consider the data simulated in accordance with formula: $y_j = 10\sqrt{j+10} + 2\text{random}_j$, where $j = 1 \ldots N$, random_j—pseudorandom numbers, evenly distributed in the segment [0,1]. Let us set N = 150 and model 10 outliers at points $j = 6\ldots10$ and $j = 140\ldots144$ (see **Figure 8**). Further, let us set $\sigma_{max} = 0.6$, MINOBS = 10 and search for outliers as described above with L = 130. If $n = 1$, we get $N_{out} = 88$; if $n = 2$, we get $N_{out} = 33$; if $n = 3$, we get $N_{out} = 17$; if $n = 4$, we get $N_{out} = 10$. Thus, a suitable trend approximation turned out to be a fourth degree polynomial. **Figure 9** shows values for fourth degree polynomial fitted to the data set on

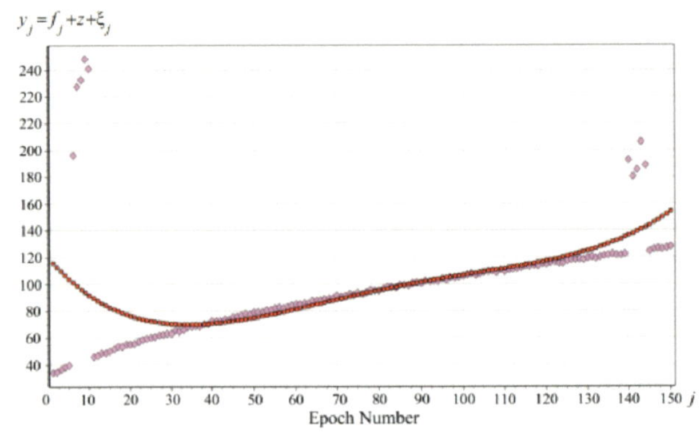

Figure 8.
Data simulated using: $y_j = 10\sqrt{j+10} + 2\text{random}_j$ *and fourth degree polynomial after the first iteration.*

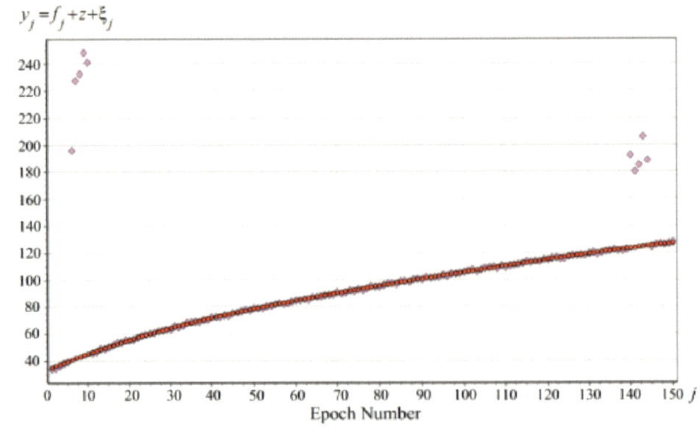

Figure 9.
Simulated data and fourth degree polynomial after the eighth iteration.

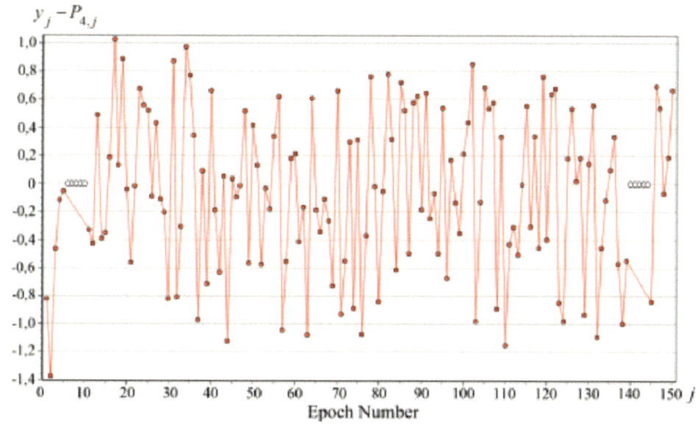

Figure 10.
The differences $\hat{y}_j = y_j - P_{4,j}(\vec{a})$, approximation of unknown trend with fourth degree polynomial.

the eighth iteration after the convergence discussed above is reached. In **Figure 10**, the differences $\hat{y}_j = y_j - P_{4,j}(\vec{a})$ are plotted. Positions of detected outliers are marked with white circles. Note that the trend was modeled by a function that does not belong to the class of power polynomials.

7.5 Convergence of iterations in trend search algorithm

This section explains the convergence of the iterations described in the trend search algorithm (see previous section).

Assertion 7. *The SD sequence $\sigma_{Y^{(s)}_{L,\min}}$ of values calculated in the trend search algorithm monotonically decreases at s = 1, 2, ... :*

$$\sigma_{Y^{(s)}_{L,\min}} \geq \sigma_{Y^{(s+1)}_{L,\min}} \geq \dots \tag{54}$$

and therefore converged.

Proof. We start our consideration with Step 3 and sth iteration, s = 1, 2, ... In Step 3, for the sequence $\left\{\hat{y}^{(s)}_j\right\}_{j=1}^{N}$, we find a minimizing set of length L: $Y^{(s)}_{L,\min} = \left\{\hat{y}^{(s)}_{j^{(s)}_1}, \dots, \hat{y}^{(s)}_{j^{(s)}_L}\right\}$. Write the expressions for the mean and the square of SD corresponding to this set. We have due to Eqs. (40) and (41):

$$z_{Y^{(s)}_{L,\min}} = L^{-1} \sum_{j \in \left\{j^{(s)}_1, \dots, j^{(s)}_L\right\}} \hat{y}^{(s)}_j, \tag{55}$$

$$\sigma^2_{Y^{(s)}_{L,\min}} = (L-1)^{-1} \sum_{j \in \left\{j^{(s)}_1, \dots, j^{(s)}_L\right\}} \left(\hat{y}^{(s)}_j - z_{Y^{(s)}_{L,\min}}\right)^2. \tag{56}$$

Transform the expression on the right side of Eq. (56). Substitution here instead of $\hat{y}^{(s)}_j$ expression in Eq. (52) gives:

$$\sigma^2_{Y^{(s)}_{L,\min}} = (L-1)^{-1} \sum_{j \in \left\{j^{(s)}_1, \dots, j^{(s)}_L\right\}} \left(y_j - P_{n,j}\left(\vec{a}^{(s)}\right) - z_{Y^{(s)}_{L,\min}}\right)^2.$$

Using the $\text{flag}_j^{(s)}$ defined in Eq. (53), rewrite the last equality as follows:

$$\sigma^2_{Y^{(s)}_{L,\min}} = (L-1)^{-1}\sum_{j=1}^{N}\left(y_j - P_{n,j}\left(\vec{a}^{(s)}\right) - z_{Y^{(s)}_{L,\min}}\right)^2 \cdot \text{flag}_j^{(s)} =$$
$$= (L-1)^{-1}\sum_{j=1}^{N}\left(y_j - P_{n,j}\left(\vec{b}\right)\right)^2 \cdot \text{flag}_j^{(s)} = (L-1)^{-1}\Phi^{(s)}\left(\vec{b}\right). \quad (57)$$

Here we introduce the designation $\vec{b} = \left\{a_0^{(s)} + z_{Y^{(s)}_{L,\min}}, a_1^{(s)}, \dots, a_n^{(s)}\right\}$ and take into account the definition of $P_{n,j}\left(\vec{a}\right)$ expressed by Eq. (37) and the definition of functional $\Phi^{(s)}\left(\vec{a}\right)$ given in Eq. (51). From Eq. (57), we get:

$$\sigma^2_{Y^{(s)}_{L,\min}} = (L-1)^{-1}\Phi^{(s)}\left(\vec{b}\right) \quad (58)$$

We transit to Step 2; *s* is incremented by 1. We find a vector $\vec{a}^{(s+1)}$ of polynomial coefficients, which minimizes the functional $\Phi^{(s)}\left(\vec{a}\right)$:

$$\vec{a}^{(s+1)} = \arg\min_{\vec{a}} \Phi^{(s)}\left(\vec{a}\right).$$

Thus,

$$\Phi^{(s)}\left(\vec{a}^{(s+1)}\right) = \min_{\vec{a}}\left\{\Phi^{(s)}\left(\vec{a}\right)\right\} \quad (59)$$

From the definition of the extremum of functional, it follows:

$$\Phi^{(s)}\left(\vec{a}^{(s+1)}\right) \le \Phi^{(s)}\left(\vec{b}\right). \quad (60)$$

Taking into account Eq. (58), we have:

$$\sigma^2_{Y^{(s)}_{L,\min}} \ge (L-1)^{-1}\Phi^{(s)}\left(\vec{a}^{(s+1)}\right). \quad (61)$$

The extremum condition (one of n + 1) of functional $\Phi^{(s)}\left(\vec{a}\right)$ is as follows:

$$\frac{\partial}{\partial a_0}\Phi^{(s)}\left(\vec{a}\right)\Big|_{\vec{a}=\vec{a}^{(s+1)}} = 0.$$

From here, we derive:

$$\sum_{j=1}^{N}\left(y_j - P_{n,j}\left(\vec{a}^{(s+1)}\right)\right)\cdot \text{flag}_j^{(s)} = 0.$$

Taking into account the designation for $\hat{y}_j^{(s+1)}$ given by Eq. (52), we can write this equality in the form.

$$\sum_{j=1}^{N}\hat{y}_j^{(s+1)}\cdot \text{flag}_j^{(s)} = 0 \text{ or } \sum_{j\in\left\{j_1^{(s)},\dots,j_L^{(s)}\right\}}\hat{y}_j^{(s+1)} = 0. \quad (62)$$

Consider the set $Y_L^{(s+1)} = \left\{\hat{y}_{j_1^{(s)}}^{(s+1)}, \dots, \hat{y}_{j_L^{(s)}}^{(s+1)}\right\}$. The mean and SD values for it are calculated using formulas (38) and (39). Taking into account Eq. (62), we get:

$$z_{Y_L^{(s+1)}} = L^{-1} \sum_{j \in \left\{j_1^{(s)}, \dots, j_L^{(s)}\right\}} \hat{y}_j^{(s+1)} = 0,$$

and

$$\sigma^2_{Y_L^{(s+1)}} = (L-1)^{-1} \sum_{j \in \left\{j_1^{(s)}, \dots, j_L^{(s)}\right\}} \left(\hat{y}_j^{(s+1)} - z_{Y_L^{(s+1)}}\right)^2 = (L-1)^{-1} \sum_{j \in \left\{j_1^{(s)}, \dots, j_L^{(s)}\right\}} \left(\hat{y}_j^{(s+1)}\right)^2 =$$

$$= (L-1)^{-1} \sum_{j=1}^{N} \left(y_j - P_{n,j}\left(\vec{a}^{(s+1)}\right)\right)^2 \cdot flag_j^{(s)} = (L-1)^{-1} \Phi^{(s)}\left(\vec{a}^{(s+1)}\right).$$

Thus,

$$\sigma^2_{Y_L^{(s+1)}} = (L-1)^{-1} \Phi^{(s)}\left(\vec{a}^{(s+1)}\right). \tag{63}$$

We transit to Step 3. In this step, for sequence $\left\{\hat{y}_j^{(s+1)}\right\}_{j=1}^{N}$, we find a minimizing set of length L: $Y_{L,\min}^{(s+1)} = \left\{\hat{y}_{j_1^{(s+1)}}^{(s+1)}, \dots, \hat{y}_{j_L^{(s+1)}}^{(s+1)}\right\}$. The SD square $\sigma^2_{Y_{L,\min}^{(s+1)}}$ for this set does not exceed the same magnitude for any other set, particularly for $Y_L^{(s+1)} = \left\{\hat{y}_{j_1^{(s)}}^{(s+1)}, \dots, \hat{y}_{j_L^{(s)}}^{(s+1)}\right\}$:

$$\sigma^2_{Y_L^{(s+1)}} \geq \sigma^2_{Y_{L,\min}^{(s+1)}}. \tag{64}$$

Finally, from Eqs. (61) and (63) and the last inequality (64), we get

$$\sigma^2_{Y_{L,\min}^{(s)}} \geq (L-1)^{-1} \Phi^{(s)}\left(\vec{a}^{(s+1)}\right) = \sigma^2_{Y_L^{(s+1)}} \geq \sigma^2_{Y_{L,\min}^{(s+1)}}$$

that is, $\sigma^2_{Y_{L,\min}^{(s)}} \geq \sigma^2_{Y_{L,\min}^{(s+1)}}$, that proves Assertion 7.

8. Detection and repair cycle slips in the Melbourne-Wübbena combination algorithm

In this and the following section, we describe cycle slip repair algorithm for observers represented in the form of Melbourne-Wübbena combination, which is often used in modern GNSS measurement data processing programs. Loss by the receiver of the carrier phase capture results in jumps in the code and phase measurement data. In the absence of jumps, as we already discussed in Section 2, the values of the combination consist of measurement noise superimposed on an unknown constant value dependent on a specified satellite-receiver pair.

In case of temporary loss of carrier phase capture by the receiver, jumps occur in a series of values representing the Melbourne-Wübbena combination. The procedure of detecting jumps and eliminating them from the values of the combination, called cycle slip repair, is one of the most important steps of preprocessing GNSS data. The main difficulty in detecting jumps is that neither the exact size of jumps nor their epochs are known. A number of algorithms, descriptions of which can be found, for example, in Refs. [3, 6, 8] are proposed for the detection of jumps. Although differing in detail, they are based on a common idea, that is, comparison of the SDs of the time series of measurement data obtained from one of the bounds of the time interval to an arbitrary moment, an epoch. If the differences of the SDs corresponding to two adjacent epochs exceed a predetermined threshold value, then a jump is declared in one of these two epochs. A drawback of similar algorithms is the frequent false detection of jumps during epochs containing rough measurements (outliers) since the values of outliers can exceed the sizes of a jump itself. On the other hand, an attempt to increase the threshold value leads to the opposite effect, an inability to recognize jumps that are small in magnitude.

Below, we propose a robust cycle slip repair algorithm that allows, more reliably than similar known algorithms, to detect jumps and determine their sizes. The proposed algorithm is based on search for so-called clusters consisting of epochs, in which the values of the combination are grouped about corresponding predefined values. Besides, this algorithm implements the above-described method of cleaning data from outliers based on the search for the optimal solution. This method, combined with Springer's algorithm used in Ref. [3], allows for the reliable determination of multiple (cascade) jumps in the Melbourne-Wübbena combination.

The Melbourne-Wübbena combination L_{M-W} can be presented in the form of the total of three terms [6]:

$$L_{M-W} = \lambda_5 n_5 + B + \nu, \tag{65}$$

where λ_5 is the formal wavelength (for all GPS satellites λ_5 = 86.16 cm and for GLONASS satellites λ_5 = 84.0 ÷ 84.36 cm); n_5 is an unknown integer, the so-called wide lane ambiguity [3]; B is the residual systematic error caused by instrumental delays relative to the specific receiver-satellite pair and, as assumed, not time-dependent; and ν is a random component. Both parts of the equality in (65) are expressed in meters. At the same time, the L_{M-W} combination is often expressed in cycles of wavelength λ_5. Let us divide both parts of Eq. (65) by λ_5. We introduce the designations $y = L_{M-W}/\lambda_5$, $\beta = B/\lambda_5$, and $\xi = \nu/\lambda_5$ and derive for each of the epochs associated with indexes $j = 1, \ldots, N$:

$$y_j = n_{5j} + \beta + \xi_j. \tag{66}$$

where n_{5j} is an unknown piecewise-constant function of a discrete argument, regarding which it is only known that it takes integer values and can undergo integer jumps; β is an unknown constant value; and ξ_j is a random variable. The problem consists in the development of an algorithm for automatic processing of the informational data y_j of form of Eq. (66), making it possible to effectively determine integer-valued jumps against the noise component and outliers.

9. Description of the proposed algorithm

Let us present the proposed algorithm as the following sequence of steps.

Step 0: We introduce parameter Δ by specifying its value as $\Delta = 2 \cdot \sigma_{max}$. We mark the indexes of epochs with function $flag_j$ as nulls: $flag_j = 0$; j = 1, …, N.

Arrange the array y_j in the ascending order and renumber it by placing the indexes in the ascending order. Denote the resulting array as Y_j. For simplicity of

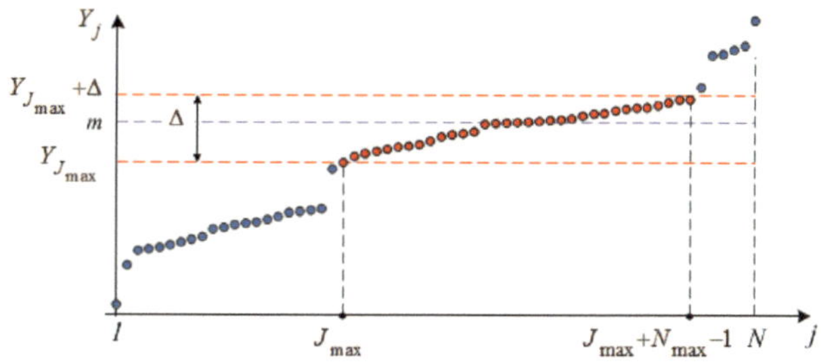

Figure 11.
Searching for the maximum of the density of values of the array Y_j.

logic, we suppose that all Y_j are different. Therefore, we have: $Y_1 < Y_2 < \dots < Y_N$. Denote the corresponding permutation of the values of function $flag_j$ as $FLAG_j$.

Step 1: On this step, we are looking for the maximum density of array values y_j. We consider the segments $[Y_j, Y_j + \Delta]$ of length Δ and look for the one that contains the largest number of Y_i values. Here, only those indexes i and j from the range of $1 \le i, j \le N$ are considered for which $FLAG_i = FLAG_j = 0$. The purpose of the ordering of the original array consists in improving the effectiveness of the maximum density search procedure. It can be shown that the number of comparisons required when searching for the above segment does not exceed 2 N in the case of the ordered array. For an unordered array, the number of comparisons is evaluated as N^2.

Step 2: Let $[Y_{J_{max}}, Y_{J_{max}} + \Delta]$ be the segment found in the previous step, containing N_{max} values of Y_j (see **Figure 11**). We calculate the mean m of these numbers Y_j.

$$m = N_{max}^{-1} \cdot \sum_{j=J_{max}}^{J_{max}+N_{max}-1} Y_j, \tag{67}$$

$$Y_j \in [Y_{J_{max}}, Y_{J_{max}} + \Delta]. \tag{68}$$

Step 3: Cluster searching. The values y_j that are clustered about the mean m found in the previous step can pertain to the epochs scattered along the time axis in chaotic order. The task of this step is to define an accumulation of such points, a so-called cluster, which we define as follows:

Definition 3. *We designate as an (m, Δ) cluster the set of points (epochs) of the time axis, the indexes j of which belong to the segment [k,l] $(1 \le k < l \le N)$ and for which the following requirements are satisfied:*

a. *all points of the segment [k, l] are marked by the same value: $flag_k = \dots = flag_l$.*

b. *at the points j= k, l of the left and right boundaries of the segment, this inequality is satisfied:*

$$|y_j - m| \le \Delta. \tag{69}$$

c. *the amount of points at which (69) is satisfied is no less than the present value of MINOBS;*

d. *the number of consecutive points in which (69) is not satisfied does not exceed the predefined value MAXGAP (e.g., 5);*

e. the value of k cannot be reduced while maintaining the requirements of a–d; and

f. the value of l cannot be incremented while maintaining the requirements of a–d.

This definition illustrates in **Figure 12**.

It is understood that for specified values of m and Δ, there might be one, several, or no (m, Δ) clusters. Searching for clusters is performed by sequential checking of the satisfaction of inequality (69) for all $j = 1, \dots, N$ for which $flag_j = 0$. Note that inequality (69) is satisfied for at least N_{max} values of j. In fact, from expressions (67) and (68), we derive the following equations:

$$Y_{J_{max}} \le m \le Y_{J_{max}} + \Delta, \tag{70}$$

and since the arrays y_j and Y_j differ only by permutation, it follows from Eq. (68) that the inequalities

$$Y_{J_{max}} \le y_j \le Y_{J_{max}} + \Delta. \tag{71}$$

are fulfilled for N_{max} index j. Inequality (69) is satisfied also for all these indexes, as follows from Eqs. (70) and (71).

If cluster was found, then we mark all points of it as 1, and then, we repeat the cluster search procedure. If even just one cluster has been found at this step, we transfer to Step 1. If not even one cluster has been found, then the search for clusters is complete and we transfer to Step 4.

Step 4: If even just one cluster has been found, we transfer to Step 5, and otherwise: (a) all points of the segment $[1; N]$ are marked as outliers and removed from further processing and (b) the operation of the algorithm terminates.

Step 5: Search for individual jumps in clusters. Let us assume that $n \ge 1$ clusters have been found:

1. $[k_1, l_1]$ is the (m_1, Δ) cluster

...

n. $[k_n, l_n]$ is the (m_n, Δ) cluster.

In each of the clusters that are found, outliers and 1-size jumps are possible. This follows immediately from the inequalities in (69) and the preestablished value $\Delta = 1.2$.

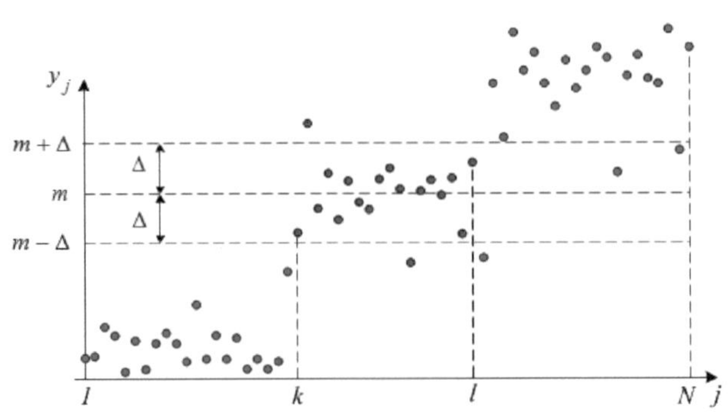

Figure 12.
Epochs with indexes $k \le j \le l$ form (m, Δ) cluster.

Substep 5.1: In detecting a 1-size jump in a cluster, we use modified Springer algorithm (see Refs. [9, 10]) combined with the proposed in Section 3 algorithm that executes a search for the optimal solution with a minimum quantity of defective data.

Substep 5.2: We find all epochs J_p and values $\Delta n_{5,J_p} \triangleq n_{5,J_p} - n_{5,J_p-1}$ $(p = 1, \dots, n)$ of the jumps inside the clusters. The possible values for $\Delta n_{5,J_p}$ are $0, \pm 1$.

Substep 5.3: We repair the data by the value of each found jump, using the formula

$$y_j^{(p)} = \begin{cases} y_j^{(p-1)}; & j < J_p \\ y_j^{(p-1)} - \Delta n_{5,J_p}; & J_p \le j < N \end{cases}; \quad p = 1, .., n, \tag{72}$$

where $y_j^{(0)} = y_j$.

Substep 5.4: We rename: $y_j = y_j^{(n)}$.

Step 6: Marking of points outside clusters. All points outside of the found clusters (if any) are marked as outliers.

Step 7: Ordering of clusters. We renumber the clusters so that they are placed left to right on the time axis. For the ordered set of clusters $[k_p, l_p], p = 1, \dots, n$, the conditions $1 \le k_1 < l_1 < k_2 < l_2 < \dots < k_n < l_n \le N$ are satisfied.

Step 8: Data screening within clusters and improving the mean values of y_j.

Substep 8.1: In accordance with the algorithm proposed in Section 3, we perform screening from outliers in each of the n clusters.

Substep 8.2: For each of the clusters cleaned of outliers, we determine the modified mean values m_p^*.

Step 9: Jumps between clusters. It follows from the description presented above that the remaining jumps in the data y_j are on boundaries between clusters. If only one cluster is found, the algorithm is terminated: no additional jumps are detected, and the data are cleaned of outliers. If more than one cluster is found ($n > 1$), then the epochs $j_1, \dots, j_{n-1}$ of the jumps will be the coordinates of the left-hand boundaries of clusters, beginning from the second: $j_1 = k_2, \dots, j_{n-1} = k_n$. The values of jumps $\Delta n_{5,j_p}$ are found in this case as rounded to the nearest integer of the difference of mean values of adjacent clusters:

$$\Delta n_{5,j_p} = NINT\left(m_{p+1}^* - m_p^*\right), p = 1, \dots, n\text{–}1. \tag{73}$$

Step 10: Repair data. We delete the jumps between clusters using formula analogous to (72):

$$y_j^{(p)} = \begin{cases} y_j^{(p-1)}; & j < j_p \\ y_j^{(p-1)} - \Delta n_{5,J_p}; & j_p \le j < N \end{cases}; \quad p = 1, \dots, n-1, \tag{74}$$

where $y_j^{(0)} = y_j$.

We rename: $y_j = y_j^{(n-1)}$. The algorithm is terminated.

10. Numerical calculations using algorithms presented in Sections 8 and 9

We present here the results of testing the proposed algorithm using real data obtained by the JOZ2 station, which is part of the IGS network [2]. These data are included in the distribution set of the installation software package [3]. Testing was carried out for data obtained from GPS satellite with number PRN = 13 for 2010, day 207. **Figure 13** shows the Melbourne-Wübbena combination values in the

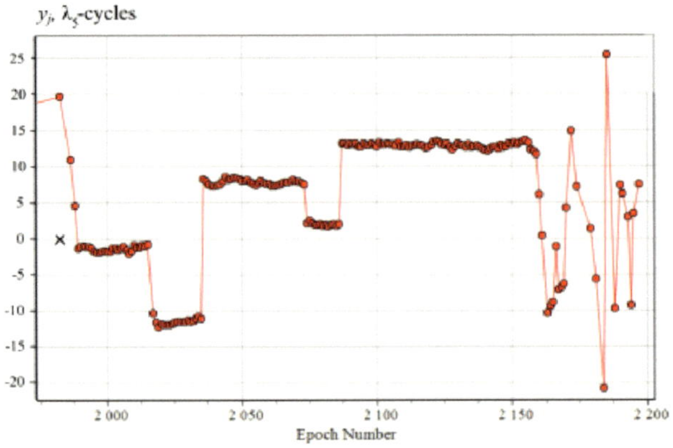

Figure 13.
Values of the Melbourne-Wübbena combination for the JOZ2 station (PRN = 13 for 2010, day of year = 207).

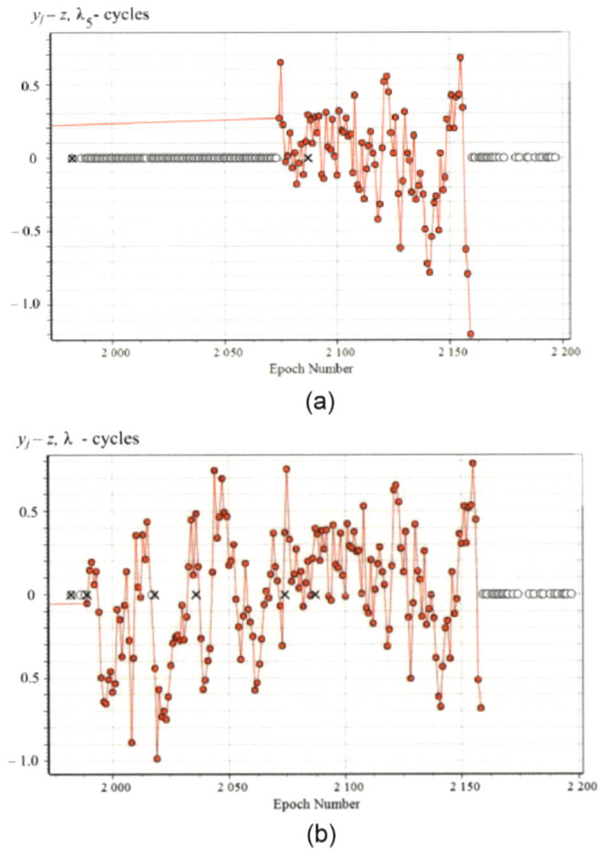

Figure 14.
(a) Deviations of the values of the Melbourne-Wübbena combination from the mean after detection and elimination of jumps from the data using the algorithm from Ref. [3]. (b) Deviations of the values of the Melbourne-Wübbena combination from the mean after detection and elimination of jumps from the data using the proposed algorithm in Section 9.

107 min time interval (N = 215). The number j of time epochs counted from the beginning of 24-h day with interval of 30 seconds is plotted along the horizontal axis. The combination values y_j expressed in cycles with wavelength λ_5 are plotted along the vertical axis.

Figure 14a and **b** presents the values of the deviations from the mean value z of data after cycle slip repair procedure, by using the algorithm applied in Ref. [3] (see **Figure 14a**) and the proposed algorithm (**Figure 14b**). The values y_j–z in cycles of λ_5 are plotted along the vertical axes, and the number j of epochs is plotted along the horizontal axis. Epochs in which the measurement data were rejected are marked by light circles. In the first case (see **Figure 14a**), 111 of the conducted measurements or 51% of the total number of data was discarded. In the second case (see **Figure 14b**), 29 of these measurements were rejected (13%), which are almost 38% less than in the previous computation. The epochs of detected jumps are marked by daggers.

11. Conclusion

This chapter presents several effective and stable algorithms for processing data received from GNSS receivers. These data form the basis of almost all engineering applications in the field of computational geo-dynamics and navigation and cadastral survey and in numerous fundamental research works as well. The accuracy of the results obtained is significantly influenced by the quality of the data used in the calculations. In particular, the presence of rough measurements (outliers) in the observation data can significantly reduce the accuracy of the calculations carried out. One of the tasks at the preliminary stage of data processing is reliable detection and removal of rough measurements from the series of measured data with minimum amount of rejected data. The so-called optimal solution, introduced in the chapter, made it possible to detect and eliminate outliers from observed data minimizing the number of rejected measurements. In addition, it is assumed that the data may contain a trend as an unknown function of time. The strategy for determining of the trend is depending on the physical process in question under an assumption that the trend is a continuous function of time. The efficiency of the search is definitely influenced by the choice of the function class from which the trend is searched. In this chapter, we considered the class of power polynomials, but in some cases, this choice may not lead to the expected result. It may require, for example, a class of trigonometric functions to find a suitable trend. The automatic search for the best functional class, together with the strategy of effectively finding an unknown trend, against the background of random noise and outliers, is a complex task for future research.

Author details

Igor V. Bezmenov
"VNIIFTRI", Mendeleevo, Moscow region, Russian Federation

*Address all correspondence to: bezmenov@vniiftri.ru

References

[1] Perov AI, Kharisov VN. GLONASS: Principles of Construction and Operation. 4th ed. Moscow: Radiotekhnika; 2010

[2] International GNSS Service. Available from: wwv.igs.org/network [Accessed: 4 April 2018]

[3] Dach R, Andritsch F, Arnold D, et al. In: Dach R et al, editors. Bernese GPS Software, Ver. 5.2. Astronomical Institute, University of Bern; 2015. DOI: 10.7892/boris.72297. Available from: www.bernese.unibe.ch [Accessed: 20 May 2020]

[4] Melbourne WG. The case for ranging in GPS based geodetic systems. In: Goad C, editor. Proceedings of the 1st International Symposium on Precise Positioning with the Global Positioning System. Rockville, Maryland: US Department of Commerce; 1985. pp. 373-386

[5] Wubbena G. Software developments for geodetic positioning with GPS using TI 4100 code and carrier measurements. In: Goad C, editor. Proceedings First International Symposium on Precise Positioning with the Global Positioning System. Rockville, Maryland: US Department of Commerce; 1985. pp. 403-412

[6] Sanz Subirana J, Juan Zornoza M, Hernández-Pajares M. Detector Based in Code and Carrier Phase Data: The Melburne–Wübbena Combination. ESA Navipedia. Available from: https://gssc.esa.int/navipedia/index.php/Detector_based_in_code_and_carrier_phase_data:_The_Melboume-Wübbena_combination [Accessed: 24 January 2019]

[7] Bezmenov IV, Naumov AV, Pasynok SL. An effective algorithm for elimination of outliers from data measurements of global navigation satellite systems. Measurement Techniques. 2018;**61**:878-884. DOI: 10.1007/s11018-018-1518-y

[8] Blewitt G. An automatic editing algorithms for GPS data. Geophysical Research Letters. 1990;**17**(3):199-202

[9] Springer TA. Modeling and validating orbits and clocks using the global positioning system. In: Geodätisch-Geophysikalische Arbeiten in der Schweiz. Vol. 60. Zurich: Institute. Geodesy and Photogrammetry, ETH Zurich; 2000

[10] Bezmenov IV, Blinov IY, Naumov AV, et al. An algorithm for cycle-slip detection in a Melbourne–Wübbena combination formed of code and carrier phase GNSS measurements. Measurement Techniques. 2019;**62**: 415-421. DOI: 10.1007/s11018-019-01639-5

Chapter 10

Analysis of Spatiotemporal Variability of Surface Temperature of Okhotsk Sea and Adjacent Waters Using Satellite Data

Dmitry Lozhkin

Abstract

The chapter is divided into 5 parts. The first part describes what the satellite data is and describes the levels of its processing. In the second part, attention is paid to sea surface temperature anomalies, the conditions for the appearance of the most significant anomalies and their influence on the behavior and survival of aquatic organisms are described. The third part is devoted to calculating linear trends in ocean surface temperature from a 20-year series of satellite data. It is shown that the heat content of the surface layer of Okhotsk Sea decreases, most significantly in its northern and western parts. This trend is especially pronounced in the spring, which may be due to a decrease in ice cover and a more significant cooling of the waters due to winter convection. In the fourth part, periodic fluctuations in the temperature of the surface of the ocean are considered. It is demonstrated how, using the calculated trend and several basic harmonics, one can try to predict the temperature next year. And the last part concludes the chapter.

Keywords: satellite data, sea surface temperature, Tsushima current, Okhotsk Sea, regional effects of global warming, SST anomalies, trends, harmonic, cyclicity

1. Introduction

The study of the thermal regime of various water areas is one of the most important oceanological problems, since the spatio-temporal variability of water temperature reflects complex processes of formation, transformation and dynamics of water masses. In addition, temperature is one of the key parameters that determine the conditions for the existence and development of most species of aquatic organisms; therefore, the study of this problem is also of key importance for hydrobiology. The zoning of water areas by the nature of temperature conditions, as well as their forecasting, taking into account the peculiarities of seasonal and interannual variability, is an important scientific task, which also has a pronounced applied aspect associated with the fact that accumulations of some species of commercial fish are confined to the zones of separation of water masses with different characteristics.

Direct measurements are a traditional source of water temperature data. Their accuracy depends only on the accuracy of the device, and the discreteness depends on the specific task (in some cases it can reach a fraction of a second). They can also be used to construct vertical profiles with high vertical resolution. However, recently the number of expeditions has been steadily declining, and the data obtained through direct measurements are point and irregular and are not suitable for studying large-scale phenomena or for obtaining data from hard-to-reach places.

Satellite data, on the other hand, are regular and allow covering the entire water area of the studied basin. Therefore, they are very good sources for studying seasonal and interannual variations in the temperature of the surface layer (the thickness of which ranges from 1 to 10 meters). The disadvantage of these data is the strong dependence of accuracy on cloudiness (especially in the infrared and visible ranges). Ideally, they should be regularly compared with direct measurements in order to identify errors in data interpretation.

For scientific purposes, data are usually used from satellites located either in a geosynchronous (in a particular case, geostationary orbit) or in a heliosynchronous orbit. The advantage of the geostationary orbit is obvious: you can get a picture of the same area with discreteness of up to half an hour, and after pointing to the satellite, a constant correction of the antenna position is not required. It is very useful for telecommunication systems as well as for obtaining meteorological data. The disadvantage is the lack of coverage at polar latitudes.

This disadvantage is easily eliminated by satellites in a polar sun-synchronous orbit. A sun-synchronous orbit is important to science because it keeps the angle of incidence of sunlight on the Earth's surface more or less constant, although the angle will change with the change of seasons. Without a sun-synchronous orbit, you would have to account for changes in shadow and lighting angles, making it difficult to track changes over time. It would simply be impossible to collect the information needed to study climate change.

In 1997, a TeraScan satellite receiving station was installed at SakhNIRO, with the help of which data are received from the NOAA, Metop, Aqua and Terra satellites in a polar sun-synchronous orbit (**Figure 1** [1]). At the moment, a 21-year series of satellite data on the temperature of the surface of the Sea of Okhotsk and adjacent waters has been accumulated, which makes it possible to analyze the interannual variations of this parameter.

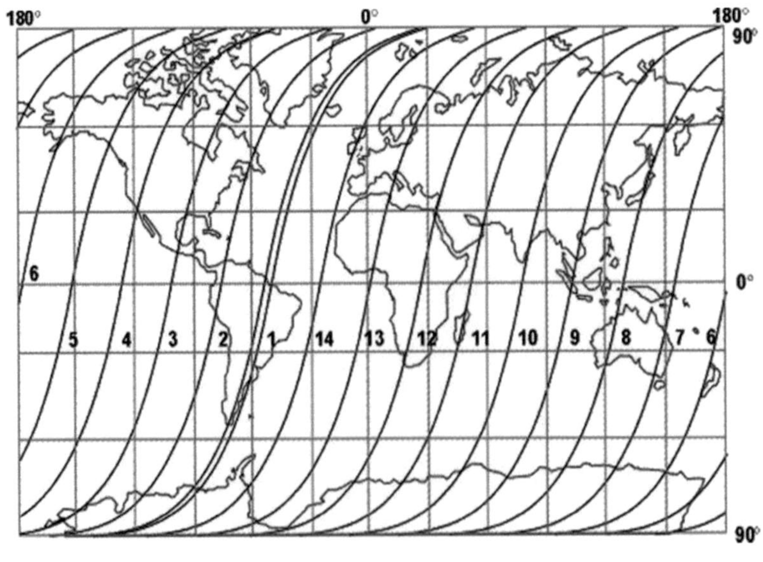

Figure 1.
Example of a sun-synchronous orbit for a NOAA satellite [1].

This chapter is organized as follows:

- Section 2 discusses satellite data
- Section 3 discusses Sea Surface Temperature (SST) anomalies
- Section 4 analyzes long-term data series using linear regression
- Section 5 presents results on periodic SST fluctuations
- Section 6 concludes the chapter.

2. Satellite data

What is satellite data? An electromagnetic (EM) signal with certain characteristics leaves the sea and contains information about the primary observable quantities, which are light, radiation temperature, roughness and sea level. This signal travels through the atmosphere, where it can be distorted, and where noise is added to the signal before it is received by a sensor that registers certain properties of the radiation and converts each measurement into a digital signal to be encoded and sent back to earth. The sensor geometry limits each individual observation to a specific instantaneous field of view.

In order to convert a digital signal received at a ground station into a useful one for scientific measurements with a certain accuracy and quality, the process of obtaining data by the sensor should be numerically transformed using knowledge about signal processing technology, physical and technical properties of the sensor, the physics of the atmosphere, and the interaction of EM radiation with the ocean and near-surface processes in the ocean. The process of extracting data from a unique observation location in space is at the heart of the satellite data feature.

Table 1 summarizes 5 levels of satellite data processing.

Using the TeraScan software, raw data received from satellites is automatically calibrated and atmospheric correction, as well as georeferencing of data and projection onto a standard map (selection of parameters, projections, etc., is performed

Level	Product description
0	Raw satellite data in binary format.
1	Image in measurement coordinates. Satellite data of individual calibrated channels.
1,5 (or 1A)	In special cases, a Level 1 product with atmospheric correction applied.
2	Atmospheric corrected and calibrated product containing upstream values or directly variable ocean response with reference to geographic coordinates, but most often not mapped.
3	Composite images of the obtained ocean variables transferred to a standard map, obtained by averaging over space and time of several flights in the form of a level 2 product. This product contains data from only one sensor and may have gaps.
4	An image representing the ocean variable averaged within each cell of the spatio-temporal grid, for the creation of which the gaps in the level 3 product data are filled with data analysis methods, incl. Interpolation. In the course of the analysis, it is possible to use several level 2 or 3 products from different sensors, and it is also possible to use data from field observations or model calculations.

Table 1.
A summary of various satellite data processing levels.

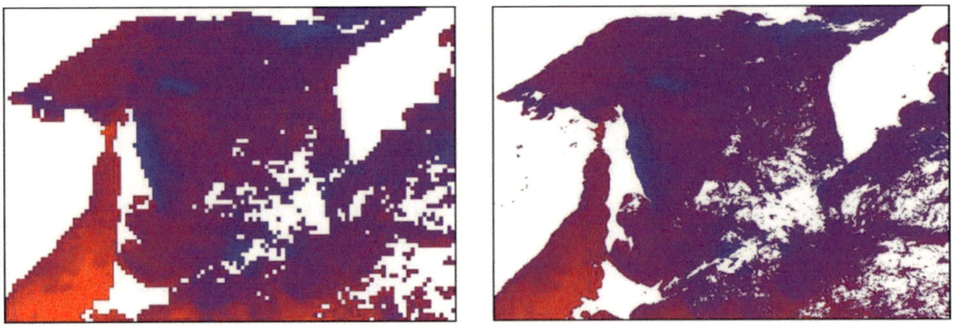

Figure 2.
An example of a ten-day average image with a spatial resolution of 0.25 × 0.25 degrees (left) and 2 × 2 km (right) [2].

once by the operator, after which it is performed in automatic mode). Then data with different spatial resolution (the maximum resolution for a set of satellites is 2 km, images with a spatial resolution of 250 m can be obtained on separate channels) are entered into the database, where the operator manually corrects each image (eliminates gross errors that most often appear in the case of the presence of semi-transparent clouds). After that, daily data averaging is performed. As the data accumulate, ten-day and monthly averaging is also performed (as shown in **Figure 2** [2]), as well as the calculation of anomalies (deviations from the long-term average value in a specified decade or month, obtained over a number of previous years).

The resulting data set should be carefully analyzed. Determination of the characteristics of unidirectional trends, as well as the amplitudes and phases of cyclic components and the possibility of using them to predict thermal conditions in the next season constituting the content of this study.

3. SST anomalies

There are two types of SST anomalies as for any parameter distributed in space and time. The first is the sharp deviation of the specified parameter from the mean values in the neighboring areas. Anomalies of this kind can be regular and even present on average long-term maps (**Figure 3** [3]). Typical examples from the point of view of SST are upwelling zones (quasi-stationary), mouths of large rivers, as well as zones of influence of warm and cold currents. Some anomalies of this nature may be unstable and dependent on weather conditions. Such anomalies determine the behavior of aquatic organisms, since each species has its own conditions of existence and development (some species need colder water, and some are more comfortable in warmer water).

The second type of SST anomaly is a deviation from the multiyear average. This type of anomaly depends entirely on the current conditions and reflects the peculiarities of the distribution of the parameter in a given year. From the point of view of SST, this may mean more or less intense heating or cooling than usual. The appearance of such anomalies can lead to a change in the timing of the fish's approach to spawning (**Figure 3** [3]) or to migration to places with more favorable conditions, and in the early stages of development it can lead to death from unfavorable conditions.

Let us consider in more detail the extraordinary conditions that took place in the spring of 2011, when the greatest delay in the release of juveniles from the Sakhalin SFHs was noted. The spatial distributions of sea surface temperature anomalies observed during the period under consideration, which is strongly characterized by unusual conditions (**Figure 4** [4]), are quite remarkable.

In the third decade of May (**Figure 4A**), significant negative deviations from the norm were noted along the entire eastern coast of Sakhalin Island and in the Aniva Gulf; they were the largest (up to 3°C) in the basin of the Mordvinova Gulf. Judging by the zone of negative anomalies stretching from the Terpeniya Peninsula along the line of the depth increase, which corresponds to the core of the East Sakhalin Current [5], they were caused by the outflow of cold water and ice-cover remains from the northeastern shelf. This opinion is confirmed by a satellite image on June 2, 2011 (**Figure 3** [3]), which clearly shows the transport of drifting ice from the north to the Terpeniya Peninsula and further towards the Mordvinova Gulf. Such a situation is observed relatively rarely: usually, under the influence of the southern winds characteristic of this period of the year, the ice that is transported from the Sakhalin Gulf is blocked at the northeastern tip of the island and does not move south of 52° N. A similar ice removal was observed in spring 2005 [6]. Then, due to the unusual distribution of the surface atmospheric pressure, the northeastern

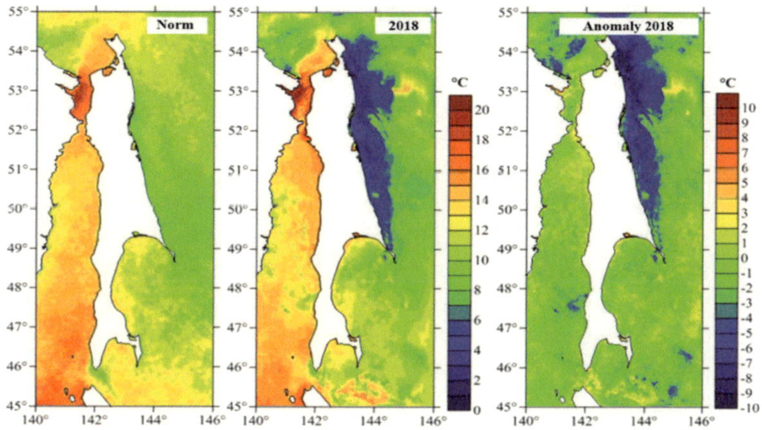

Figure 3.
An example of the long-term average distribution of SST, the actual temperature in the second decade of July 2018 and the 2018 anomaly relative to the multi-year average (these conditions led to a two-week delay in the approach of pink salmon to spawning rivers) [3].

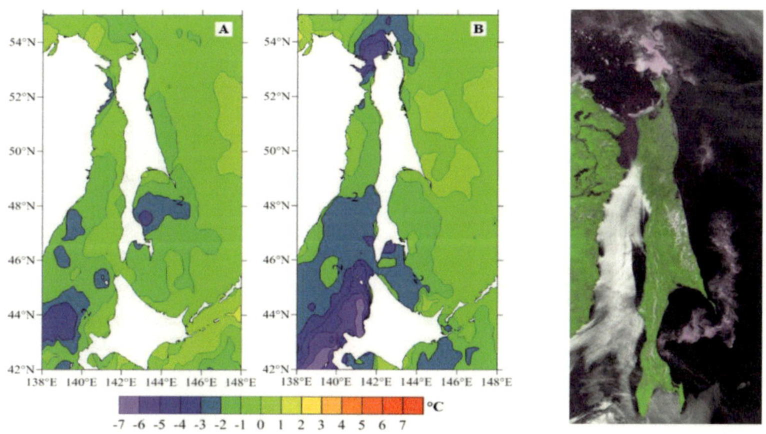

Figure 4.
Distribution of surface temperature anomalies (°C) in the third decade of may (A) and in the second decade of June (B) 2011. Ice conditions off the east coast of Sakhalin Island (right) on June 2, 2011 (snapshot of the aqua satellite). The strait of Tatary is closed by heavy clouds [4].

winds prevailed in the study region, which just contribute to the transport of drifting ice to the southeastern coast of Sakhalin Island, as well as to the estuarine regions of the Naiba and Ai rivers and to the Mordvinova Gulf.

In spring 2011, the distribution of surface atmospheric pressure also significantly differed from the characteristic one in the given period of the year. For example, in the third decade of May, the high-pressure region was located in the southern part of the Sea of Okhotsk, above the Kuril Basin (**Figure 5A** [7]). Usually this area is located in the eastern part of the sea, and air flows directed to the north are formed above this basin. When the pressure distribution has this structure, the atmospheric circulation typical for the summer monsoon is not formed; in principle, it was weakly expressed, since the pressure gradients were small and cold waters with low salinity moved freely to the south, transporting drifting ice.

Subsequently, the high-pressure region shifted from the sea to the Pacific Ocean. In particular, in the second decade of June, zonal, northeast-oriented air flows prevailed, which is not typical for the warm season (**Figure 5B** [7]).

However, it is unlikely that only the peculiarities of the synoptic conditions can explain the significant negative temperature anomalies of the sea surface in the northern part of the Sea of Japan, primarily in the main zone of the Tsushima Current influence, off the west coast of Hokkaido Island (up to 5–6°C). Another zone with approximately the same negative anomalies was noted in the zone of the influence of the Amur River runoff in the region of its mouth, in the western part of the Amur Estuary, and in a significant part of the Sakhalin Gulf, as well as on the northeastern shelf of Sakhalin Island (up to about 52° N). Pronounced negative anomalies were observed both on the southwestern and southeastern coasts of Sakhalin Island, as well as in the Aniva Gulf. It is difficult to explain the cause of such a heat deficit in the surface waters of a very wide basin: one of them could be a decrease in the intensity of the Tsushima Current, whose quasi-periodic fluctuations (a cycle with a period of 5–6 years is especially distinguished) have a significant influence on the climate of the Sakhalin-Kuril region [8]. It is also difficult to assess how often such situations can be observed which have led to a delay in the release of juveniles from Sakhalin SFHs lasting from 3 weeks to a month. Probably, due to the low temperature of coastal waters observed for a long time, there could be a mass death of juveniles of natural origin which rolled along the spawning rivers of the island.

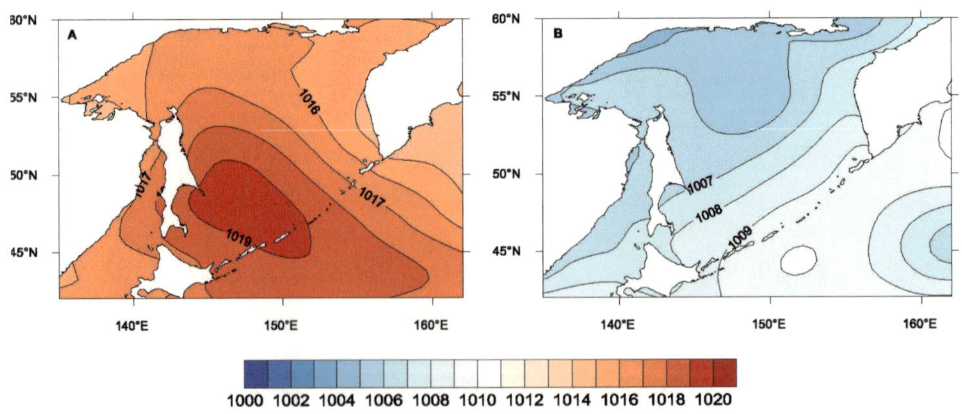

Figure 5.
Distribution of surface pressure (hPa) over the Sea of Okhotsk in the third decade of may (A) and the second decade of June (B) 2011 [7].

4. Linear trends

When analyzing long-term data series, a question arises – are there unidirectional tendencies of one or another parameter in the specified period of time? This is especially important in the context of global warming, the most noticeable results of which are an increase in winter air temperature in the Arctic latitudes and a decrease in ice cover (area of the water area covered with ice) both in the Arctic Ocean and in the Sea of Okhotsk.

In order to answer this question, in each spatial cell of the 1000 × 1022 matrix, covering the investigated water area (42–60°N and 135–163°E), the linear trend coefficients were calculated using the least squares method, meaning the temperature increase over the year ... The calculation was carried out on a 20-year series of data for each month separately, for annual averages, as well as for average temperatures for the season. The seasons for the calculations are shifted by a month relative to the calendar ones due to the specifics of the region: January–March (winter), April–June (spring), July–September (summer) and October–December (autumn).

The linear trend coefficients calculated from the full annual series in the study region are mostly negative, and there are no regions with positive values in the Sea of Okhotsk (**Figure 6** [9]). Significant cooling of the surface layer (about 1.5°C over 10 years) was observed in the northern and western parts of the sea; most clearly this process is manifested in the northern part of the sea at some distance from the coast. A less pronounced decrease (at a rate of about 0.5°C over 10 years) is observed in the Kuril Islands and in the band along the western coast of Kamchatka beyond the shelf zone, as well as in the influence zone of the Amur River outflow (Amur Estuary, Sakhalin Bay, northern shelf of Sakhalin Island).

A significant decrease (more than 0.5°C over 10 years) was found in the northern part of the Strait of Tatary; further southward, the cooling rate of surface waters decreases and, south of 45° S, a slight increase in the temperature of the surface waters in the Sea of Japan was recorded. The warming process is more pronounced in the northwestern part of the Pacific Ocean, i.e., in the southeastern part of the study region.

Let us now consider the results of calculating the parameters of the linear trend in different seasons. In winter, there is not enough data for a reliable calculation in the northwestern part of the sea (the "Sea of Okhotsk fridge"), on the northern shelf of the sea, and on the northeastern shelf of Sakhalin. In many regions of the basin, where the ice cover is significantly influenced, the calculation is not reliable enough; nevertheless, some ideas of the trends in the thermal regime during the cold period can be put forward.

In the entire Sea of Okhotsk, especially in the northern and western parts of the sea, as well as in the Strait of Tatary of the Sea of Japan and in the regions of the Pacific Ocean adjacent to the Kuril Islands, there is a tendency for the temperature decrease in the surface layer. Signs of the opposite trend are recorded only in the southeastern part of the study region, in the Pacific Ocean, in the northern part of the Sea of Japan, and in the very southern part of the Strait of Tatary (south of 47° N).

The most pronounced decrease in the surface layer temperature is in the springtime. It covers the entire study region, including the northwestern part of the Pacific Ocean, and it is particularly significant in the northern and western parts of the Sea of Okhotsk (the rates of temperature decrease range from 1.0 to 1.5°C over 10 years and even higher in some regions).

The cooling of the surface layer in spring is the most evident consequence of a decrease in ice cover, since in the absence of the ice cover the cooling processes

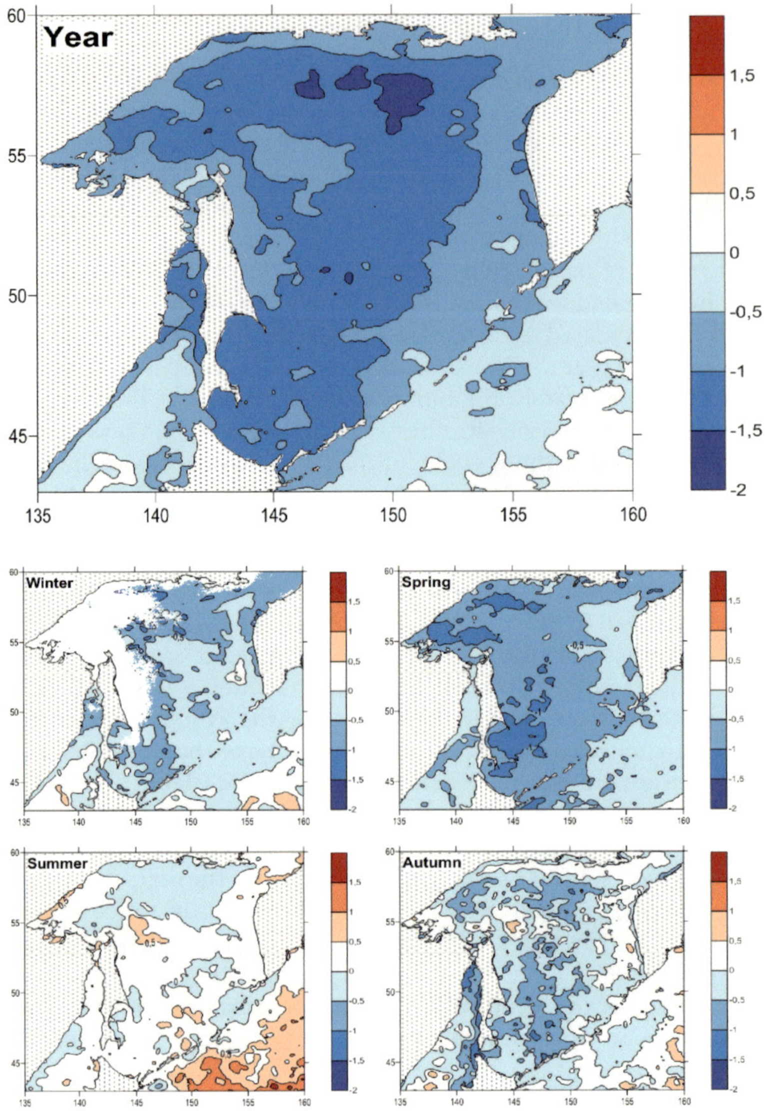

Figure 6.
Distribution of linear trend coefficients by number of average annual (above) and average seasonal values. The scale corresponds to degrees celsius over 10 years [9].

cover a layer of water thicker than in the presence of the ice cover. Undoubtedly, this has a significant impact on the climate of the region and, above all, on the weather conditions of Sakhalin, which are the most affected by this process.

While calculating the trends in sea surface temperature in individual months, it was found that the most intense cooling is observed in the month of May, especially in the northwestern part of the Sea of Okhotsk and along the entire eastern coast of Sakhalin Island from Cape Elizabeth in the north to Cape Aniva in the south. Exceptions are the western part of the Amur Estuary, which accounts for the bulk of the Amur flood, and the basin adjacent to Tauiskaya Bay. It is obvious that the hydrological regime of coastal areas experiencing the influence of the river outflow differs from that typical for the Sea of Okhotsk as a whole.

In summertime, the region is dominated by an increase in the sea surface temperature most pronounced in the northwestern part of the Pacific Ocean. In the Sea of Okhotsk, it is noted in its northwestern part, in the region of the Shantar Islands, in Sakhalin Bay, and in the Amur Estuary, to the east of the northern tip of Sakhalin Island, at the northwestern coast of the Kamchatka Peninsula, and near the Urup and Simushir islands.

In the northern part of the sea, near the Kuril Islands, on the southeastern coast of Sakhalin Island and along the Primorye coast in the Sea of Japan, there is a trend towards a decrease in temperature, although it is more moderate than in springtime. Moreover, in July, the tendency towards a decrease in the sea surface temperature still prevails in the study region; the trend changes only in the northwestern part of the Pacific Ocean. In August, in these regions, warming reaches the highest rates (about 1.5°C over 10 years); the temperature increase was noted in the central part of the Sea of Okhotsk, along the western coast of Kamchatka and off the southeastern coast of Sakhalin, in the Strait of Tatary and the Amur Estuary. In the northern part of the sea, on the northeastern shelf of Sakhalin, and in the Kuril region, the trend towards a decrease in sea water temperature continues.

In September, on the contrary, the intensity of warming in the Pacific Ocean decreases; the greatest warming rate was noted in the northern and northwestern parts of the sea (especially in the shelf zone between Ayan and Okhotsk settlements).

A diffuse pattern is observed in the fall. Relatively small negative trends prevail in the greater part of the Sea of Okhotsk; positive trends (also insignificant) are noted in the eastern part of the sea on the western shelf of Kamchatka and in a relatively narrow band along the northern coast. Smaller areas were also found in the Shantar Islands and Kashevarov Banks and in the southwestern part of the analyzed basin of the Sea of Japan.

The highest rates of the temperature decrease in the surface layer are observed in the Strait of Tatary, in its northern part, and along the western coast of Sakhalin, as well as in the Sea of Okhotsk, in the northern part of it, beyond the shelf, off the southeastern coast of Sakhalin, and also in the band between 145° and 150° E from the Urup and Iturup islands to the Kashevarov Banks.

It was found in the calculations for individual months that a positive trend prevailed in October, and the trend reversed in November and December.

In the work [10], it was found that, in the Sea of Okhotsk and in the region of the Kuril Islands, negative trends were also noted in the spring months; the largest positive trends were recorded in October. Similarity results become very interesting if we take into account the differences in the studied time periods and spatial characteristics of the regions in which the trends were calculated. In the Sea of Japan, no negative trends were identified in this work.

As a result of the analysis of data on the surface temperature of the Sea of Okhotsk over a 20-year period (1998–2017), it was found that the global climate change in this basin caused a decrease in the ice cover and a decrease in the temperature of the upper water layer during the winter–spring period. The negative trends in temperature in the spring in the northern and western parts of the study region, as well as in the Strait of Tatary of the Sea of Japan (from 0.5° to 1.5°C over 10 years), are especially large. This effect in the reduction of the ice cover, both in time and in space, is the most evident and can be explained by an increase in the depth of winter convection. The predominance of a decrease in sea surface temperature, although less pronounced, was also noted in winter and autumn, and generally throughout the year.

In summer, the region is dominated by an increase in sea surface temperature, most pronounced in the northwestern part of the Pacific Ocean. In the Sea of Okhotsk, it was noted in its northwestern part, east of the northern tip of Sakhalin Island, off the northwestern coast of the Kamchatka Peninsula, as well as in some other regions. Moreover, in July, the main role belongs to the processes of cooling of the surface layer; the change in the trend occurs in August and continues in September–October.

These processes play a significant role in the climate variations in the Sea of Okhotsk; in particular, a decrease in the temperature of sea water is noted in the coastal waters of Sakhalin Island. In addition to the weather conditions, the results are important for studying the habitat conditions of commercial fish species and invertebrates in the basin, which is of great fishery importance.

5. Periodic fluctuations

If you look closely at the graph of the course of the mean monthly SST (**Figure 7** [11]), then in addition to seasonal fluctuations, you can notice interannual variations. They are especially pronounced in August, when the mean monthly water temperature reaches its maximum value, and are expressed in the modulation of the annual harmonic. Moreover, these oscillations are of a quasiperiodic nature, most likely associated with certain phenomena in the atmosphere and hydrosphere.

Any periodic oscillations can be described by knowing their amplitude, phase and period. Using the least squares method, you can find the corresponding amplitude and phase for each selected period. Thus, it is possible to establish a kind of "influence zones" of harmonics with a certain period, i.e. areas in which the amplitude of one or another harmonic exceeds a certain threshold value. And in order not to interfere in the calculations with long-term components (the period of which exceeds half the length of the series), the trend obtained according to the method described in the previous section was subtracted from the initial data.

The distribution of the amplitudes of temperature fluctuations in the studied region is rather complex. Having determined in each spatial cell the period corresponding to the largest amplitude and displaying the obtained data on the screen, we have established several main periods that play a significant role in interannual SST variations in most cases. For a series of 21 years, it is not entirely correct to calculate fluctuations with a period exceeding 11 years. Short-period fluctuations are unstable and generally have little information. Thus, the spatial distributions of harmonic amplitudes with a period from 3 to 11 years were considered in detail.

In most of the studied water area, the main role was played by variations with a period of about 5.5–6 years (on the graphs for points 2, 5 and 8 in **Figure 8**, you can see that the largest amplitude at these points corresponds to a period of 66–68 months), the spatial distribution of the amplitude with a period of 6 years is shown in **Figure 9** [12]. The zone of its influence is the most extensive and occupies

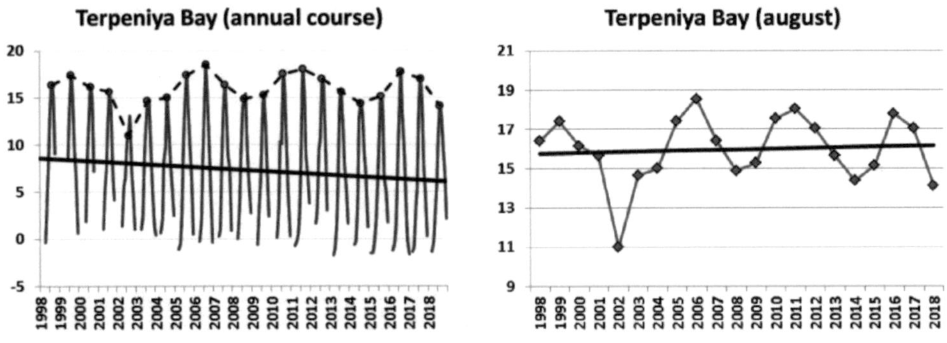

Figure 7.
Annual variation of average monthly temperature (left) and average temperature in august (right) [11].

the southern half of the Sea of Okhotsk and the northern part of the Sea of Japan, up to the Amur estuary. The amplitude of this harmonic in the zone of its influence ranges from 1 to 2°C. In the vicinity of the Kuril Islands, off the western coast of about. Hokkaido, off the northeastern coast of about. Sakhalin, the amplitude is slightly lower (from 0.5 to 1°C). Further to the north, starting from 52 N, its influence decreases and practically disappears.

In [8], based on the EOF decomposition of the sea surface temperature in the North Pacific Ocean, it was shown that an oscillation with a similar period is characteristic of the entire region influenced by the Kuroshio Current and its branch, the warm Tsushima Current. This is evidenced by the large amplitudes in the zone of the indicated currents and in **Figure 9**. Consequently, as a result of this study, it was possible to estimate the boundaries of the influence of this component in the Sea of Okhotsk, which runs parallel to the islands of the Kuril ridge and divides this basin into two practically equal parts. The entire northern Sea of Japan is significantly affected by this cycle.

As seen from **Figure 8**, at most points, one can also note peaks in the range of periods from two to three years. The highest values of the amplitude of the 3-year harmonic can be noted in the northwestern part of the Sea of Okhotsk, at a distance from the coast, as well as near the northwestern coast of Kamchatka and in the strip from 47 to 49^0 N. and from 147 to 149^0 E. in the area of the Kuril deep-water basin and in the northwestern part of the Pacific Ocean. However, the amplitudes of these oscillations are somewhat lower than those of the six-year harmonic (from 1 to 1.5°C). At a distance from these regions, the amplitude gradually decreases to zero.

The spatial distribution of the amplitudes of the cyclic component with a period of 5 years differs markedly from that considered above for a period of 6 years. The zone of its influence is noticeably narrower, it is concentrated mainly on the northern shelf of Hokkaido, in the region of the South Kuril Islands (vast waters both on

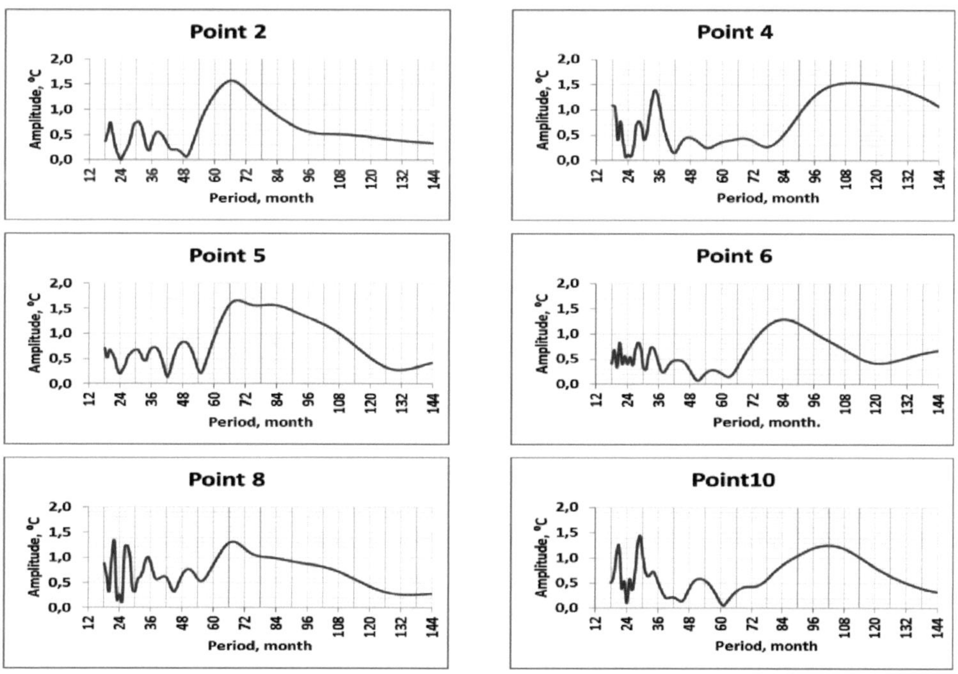

Figure 8.
Examples of graphs of the dependence of the amplitude of the harmonic (in °C) from its period (in months). The location of the points is shown in ***Figure 9*** *[12].*

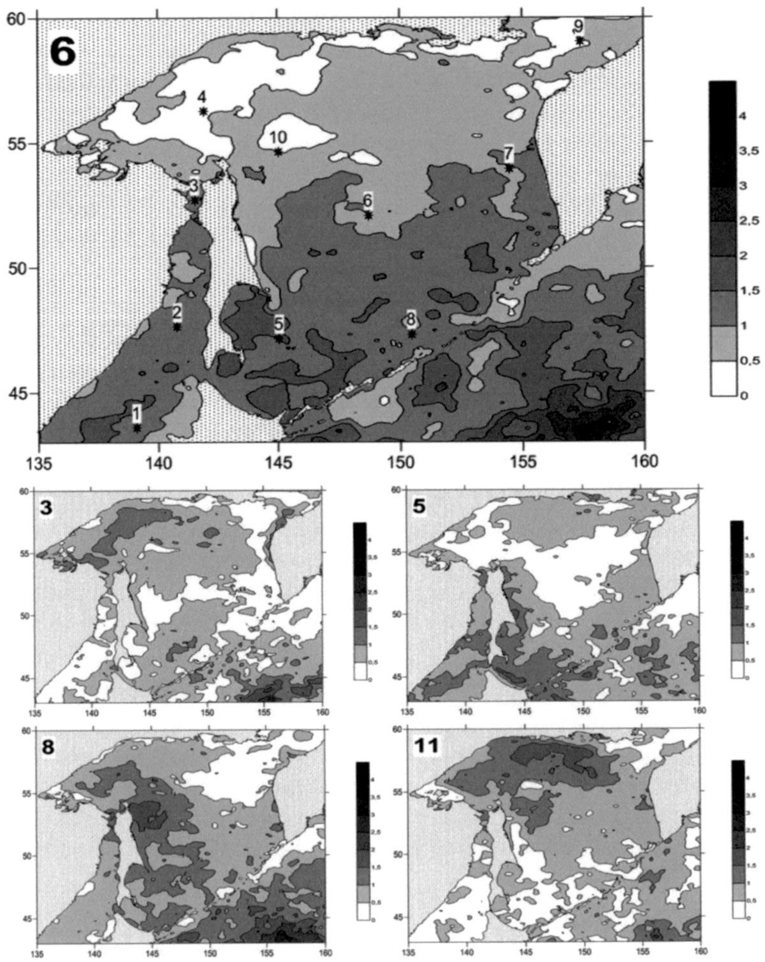

Figure 9.
Distribution of amplitudes (in °C) of interannual fluctuations in the mean monthly sea surface temperature (august) with a certain period (indicated in years in the upper left corner of the image) [12].

the Sea of Okhotsk and on the ocean sides), and, surprisingly, on the northeastern shelf of Sakhalin, where the influence of a lower frequency component was not noted. In the Tatar Strait, its role is also noticeable, but expressed to a lesser extent than the 6-year harmonic.

In the western part of the Sea of Okhotsk and the northwestern part of the Pacific Ocean, a cyclical component with a period of about 8 years is significantly manifested. The zone of influence of the Amur river runoff in summer is clearly distinguished in the spatial distribution - the Amur estuary, the southern and eastern parts of the Sakhalin Bay, the area between the Schmidt Peninsula and the Kashevarov Bank [13]. It is interesting that in the area of the Kuroshio Current manifestation, this component has large amplitudes, in the Tsushima Current zone in the Sea of Japan - insignificant, while on the northern shelf of Hokkaido and on the Sea of Okhotsk side of the Southern Kuril Islands, where the warming effect of the Soya Current affects, the amplitudes are significantly.

The lowest frequency of the considered harmonics with a period of 11 years is manifested in the northern part of the Sea of Okhotsk; in other parts of the study area, its role is insignificant. It is rather difficult to put forward a reasonable hypothesis that could explain such significant differences in the very long-term variations in SST in different parts of the same basin. It can only be assumed that due to the comparative shallowness of the northern region, the effect of the winds of the southern rumba and the greater number of sunny days than in the southern part, due to the lesser influence of cloudiness, the influence of the solar cycle is more noticeable here.

Attention is drawn to the fact how the zones of manifestation of harmonics shift with a period of 5 to 11 years. If the zone of influence of the 5-year harmonic is focused near the islands of Sakhalin, Hokkaido and the Southern Kuriles, then with an increase in the period, the region with the highest amplitude shifts clockwise (towards the northeastern coast of Sakhalin and further to the northern part of the Sea of Okhotsk). This interesting fact is also difficult to give a reasonable explanation, it requires additional study.

In work [8], a method was developed for predicting thermal conditions for a year in advance in certain areas of the studied water area (this method was also used to recover data gaps associated with the influence of cloudiness or technical reasons), which consisted in calculating the temperature in a given square in time t according to the formula:

$$T(t) = at + b + \sum_{k=1}^{N} c_k \cos(\omega_k t - \varphi_k) \quad (1)$$

where a and b are the parameters of the linear trend, ck are the amplitudes, and φk are the phases of the cyclic components (harmonics) of sea surface temperature variations. An essential feature of the method is the fact that the amplitudes and phases of the main cyclical components are calculated by the least squares method, with their periods ranging from 18 to 144 months with a step of 1 month. For each cell, a set of 3–4 harmonics was determined, which make the largest contribution to the interannual variations in SST. Since they are not orthogonal, for forecasting using formula (1), it is necessary to subtract the calculated wave from the initial series before determining the parameters of the next one in order to avoid double inclusion of coherent components (in [14, 15]) such a technique was called "Sequential spectra method").

Based on the parameters of the obtained cyclic components, a retrospective forecast of thermal conditions for the summer of 2018 was carried out (observational data for 1998–2017 were used to calculate the parameters of harmonics and a linear trend). The calculation was carried out for each spatial cell according to formula (1), taking into account the trend and four harmonic components with the highest amplitudes. For the forecast, periods from 18 to 144 months were covered. The calculation results are presented in **Figure 8** in the form of graphs of forecast curves and real variations in sea surface temperature, including the predicted values that took place in the summer of 2018. **Figure 10** shows the spatial distribution of the difference between the predicted and actual temperatures for August 2017 and 2018. The forecast was built for a year ahead along the entire previous series.

The curves, which are the sum of the trend and the first four harmonics, generally repeat the actual interannual temperature fluctuations. The correlation coefficient of the initial and predicted series at the selected points exceeds 90%. Note that even the first two harmonics in many cases provide a correlation coefficient of more than 70%. Despite the fact that 2018 was anomalous in terms of thermal conditions (the Tsushima Current and its Okhotsk branch of the Soya Current were weakened, a heat deficit was felt in the zone of influence of the Amur River runoff), and in some other areas, even in such water areas, the forecast can be considered acceptable. An example of a similar situation is given for the Tatar Strait, where the predicted value was higher than the actual one, but the general course was predicted correctly, and the error was not so great. For the northern part of the Sea of Okhotsk, the northeastern shelf of Sakhalin Island, and a number of other areas,

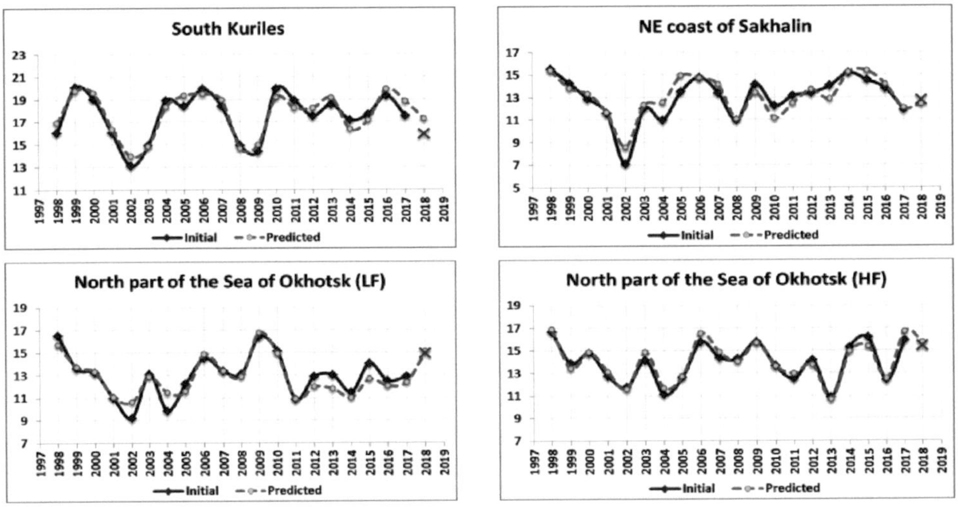

Figure 10.
Examples of temperature forecast graphs (in °C) for the next year. The dashed line shows the prognostic curve. The forecast is carried out for the period from 1998 to 2017. The actual temperature in 2018 is marked with a cross [11].

good agreement was observed between the calculated and real values of the surface layer temperature.

Let us consider some of the parameters of the graphs below. The standard deviation of the initial and predicted series correspond to each other and range from 1.5°C (Tatar Strait) to 2°C (South Kuriles). The average displacement of the predicted series relative to the initial one ranges from 0.4 to 0.6°C. The forecast error is 1.7°C in the Tatar Strait, 1.3°C near the Southern Kuriles, 0.3–0.4°C in the northern part of the studied water area.

More detailed studies devoted to predictability and the limits of applicability of the approach used will be carried out later. However, we can already say that for a significant part of the Sea of Okhotsk regions and adjacent water areas, the forecast of the surface layer temperature with a one-year lead time is quite successful, although the abnormally cold temperatures that took place in a number of areas in 2018 are rather difficult to predict.

Let us take a closer look at **Figure 11**. The forecast for August 2017 turned out to be quite successful, the discrepancy between the actual and predicted temperatures in most of the water area does not exceed ±2°C, with a standard deviation of SST of about 1.5-2°C (only in the northwestern part of the Pacific Ocean is the standard SST deviation is within 2-4°C). At the same time, the forecast for August 2018 contains a large area within which the temperature estimate was greatly overestimated (over 4°C). The map of SST anomalies for August 2018 [16] also contains areas of low temperatures (3-4°C lower than the average multiyear norm), which coincide in space with areas of unsuccessful forecast. This area is located in the zone of influence of the Tsushima Current, and the forecast inaccuracies are due to the fact that the weakening of this current occurred two years earlier than the expected date. Indeed, in **Figure 8**, we see that in the Tatar Strait and the Southern Kuriles, the distance between two neighboring SST minimums decreased to 3–4 years, while its quasiperiodic oscillations with a period of about 6 years are described in the literature [17].

As a result of the analysis of the data set on the surface temperature of the Sea of Okhotsk and adjacent waters, the main cyclical components responsible for the interannual variations of this parameter and the "zones of influence" of each

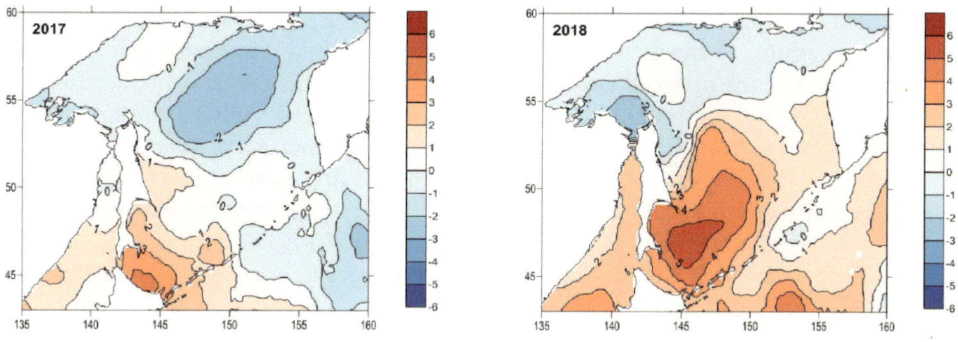

Figure 11.
Difference between predicted and actual temperatures [11].

harmonic were determined. It is shown that the main contribution to these variations comes from components with a period of about 6 years, as well as 3, 5, 8, and 11 years.

The zone of influence of the fundamental harmonic is the most extensive and occupies the southern half of the Sea of Okhotsk and the entire northern part of the Sea of Japan; its amplitude is within 1–2°C. In the vicinity of the Kuril Islands, off the western coast of Hokkaido Island, off the northeastern coast of Sakhalin Island, the amplitude is slightly lower (0.5–1°C), and in the northern part of the Sea of Okhotsk its influence is insignificant. Most likely, this component is associated with fluctuations in the Kuroshio Current and its branch, the Tsushima Current [17].

The highest values of the amplitude of the 3-year harmonic (1–1.5°C) can be noted in the northwestern part of the Sea of Okhotsk, at a distance from the coast, as well as off the northwestern coast of Kamchatka, in the region of the Kuril deep-water basin and in the northwestern parts of the Pacific Ocean.

The area of manifestation of the component with a period of 5 years is noticeably narrower, it is concentrated mainly on the northern shelf of Hokkaido, in the region of the South Kuril Islands (both from the Sea of Okhotsk and the ocean side), and on the northeastern shelf of Sakhalin. In the Tatar Strait, its role is also noticeable, but expressed to a lesser extent than the 6-year harmonic.

In the western part of the Sea of Okhotsk, in the zone of influence of the Amur River runoff, as well as in the northwestern part of the Pacific Ocean, a cyclical component with a period of about 8 years is significantly manifested.

The lowest frequency of the considered harmonics with a period of 11 years is manifested in the northern part of the Sea of Okhotsk; in other parts of the study area, its role is insignificant.

Together with the parameters of the linear trend [9], the amplitudes and phases of the main cyclical components (in each spatial cell, 4 harmonics with the highest amplitudes were used) can be used to predict thermal conditions for the next summer. The retrospective calculation for 2018 gave generally satisfactory results, despite the abnormally cold conditions of this year, noted in a number of areas of the studied water area. The possibility of predicting thermal conditions is of practical importance, primarily for assessing the conditions for the approach of Pacific salmon to spawning. And the results obtained show that in some areas of the water area, one can count on a fairly accurate forecast even for such an unstable parameter as the ocean surface temperature.

In general, the success of the forecast is influenced by how pronounced the cyclical components with a certain period in a given area. Identification of the zone of influence of various harmonics allows you to determine the boundaries of the regions in which the applicability of this method can be expected. As for the accuracy of the forecast, one should pay attention to the presence of large areas in which the modulus of the difference between the predicted temperature and the actual one was of the order of two standard deviations. This fact shows that this method does not guarantee the success of the forecast in cases where strong temperature anomalies are observed. It can only be used to obtain a primary estimate of the ocean surface temperature (or another parameter that experiences quasiperiodic oscillations) based on a sufficiently long series. The 21-year series is not enough to estimate low-frequency components (with a period of 30–50 years), which could also affect the quality of the forecast. For a better forecast, you can combine this method with an assessment of the current conditions, adjusting the forecast in the direction of increasing or decreasing temperature, depending on the current meteorological conditions.

6. Conclusion

This chapter shows how satellite data, going through all levels of processing, become useful products for both science and applied environmental prediction problems. Assessment, and later predicting the dynamics of various environmental parameters, will help reduce damage to nature and identify the degree of influence of anthropogenic factors on the future of the Earth. Based on our analysis results, it seems that the existence of global warming phenomenon has already been proven, but its influence on different parts of the planet is heterogeneous and is not traced in all seasons in the same way. In the spring, there is a tendency to a decrease in the temperature of the surface layer of the Sea of Okhotsk, this tendency continues until August, where it changes to warming condition.

Periodic fluctuations of physical parameters are also not homogeneous in space. It is possible to identify areas where the main oscillation period is 3, 5, 6, 8 and 11 years. Using the sequential spectra method gives a relatively good estimate for the next year's ocean surface temperature. However, to improve the quality of the forecast, it is necessary to improve the quality of the initial data, comparing data from various sources and filling in the gaps associated with the presence of clouds and ice cover, as well as improve the interpolation algorithms to avoid losing sight of mesoscale phenomena.

Author details

Dmitry Lozhkin
Russian Federal Research Institute of Fisheries and Oceanography, Sakhalin branch, Yuzhno-Sakhalinsk, Russian Federation

*Address all correspondence to: dima-lm@rambler.ru

References

[1] TeraScan System Training Guide (http://www.seaspace.com)

[2] Puzankov, K. L. Issledovaniye okhotomorskogo regiona pri pomoshchi sputnikovoy sistemy TeraScan [study of the Okhotsk region using the TeraScan satellite system] // Sakhalinskaya molodezh' i nauka: Materialy 1-y mezhvuz. nauch.-prakt. konf. studentov i molodykh uchen. Sakh. region (March 11-12, 1997). - 1997. – Issue 2, pp. 135-137. (In Russian).

[3] Tshay Zh. R., Shevchenko G. V. Otsenka ekstremal'nykh termicheskikh usloviy v period neresta tikhookeanskikh lososey u poberezh'ya o-va Sakhalin [assessment of extreme thermal conditions during the spawning period of Pacific salmon off the coast of Sakhalin Island] // Geodinamicheskiye protsessy i prirodnyye katastrofy, Yuzhno-Sakhalinsk, May 27-31, 2019, P. 184. (In Russian).

[4] Lozhkin DM, Tshay ZR, Shevchenko GV. Satellite monitoring of temperature conditions near the mouths of spawning Rivers in the southern part of Sakhalin Island. Izvestiya - Atmospheric and Ocean Physics. 2019;**55**(9):1166-1171

[5] Vlasova GA, Vasil'ev a S, Shevchenko GV. Prostranstvenno–Vremennaya izmenchivost' Struktury i Dinamiki Vod Okhotskogo Morya (Spatial and Temporal Variability of the Structure and Dynamics of Waters of the Sea of Okhotsk). Moscow: Nauka; 2008

[6] Shershneva OV, Shevchenko GV, Novinenko EG. Temperature conditions in the areas of juvenile release from salmon fish-breeding plants from the mouth of rivers in Sakhalin and the Iturup Island. Izv. TINRO. 2007;**150**:217-225

[7] https://rda.ucar.edu/

[8] Shershneva OV, Shevchenko GV. On the prediction of thermal conditions in the Sakhalin–Kuril region according to satellite data. Izv. TINRO. 2005;**142**:161-187

[9] Lozhkin DM, Shevchenko GV. Trends in the sea surface temperature in the Sea of Okhotsk and adjacent basins based on the satellite data in 1998-2017. Izvestiya - Atmospheric and Ocean Physics. 2019;**55**(9):1133-1137

[10] Shatilina TA, Tsitsiashvili GS, Radchenkova TV. Assessment of the statistical significance of water temperature variability in the sea of Japan and the Northwest Pacific in 1982-2007, Resursy kolyuchego kraba, perspektivy ispol'zovaniya i usloviya obitaniya: Trudy Sakhalinskogo nauchno–issledovatel'skogo instituta rybnogo khozyaistva i okeanografii (Brown king crab resources, usage prospects and habitat conditions: Transactions of the Sakhalin research Institute for Fisheries and Oceanography). Yuzhno-Sakhalinsk: SakhNIRO. 2011;**12**:180-190

[11] Lozhkin D. M., Shevchenko G. V. Lineynyye trendy i tsiklicheskiye variatsii temperatury poverkhnosti Okhotskogo morya i prilegayushchikh akvatoriy po sputnikovym dannym [linear trends and cyclical variations in the surface temperature of the Sea of Okhotsk and adjacent waters according to satellite data] // VII nauchno-prakticheskaya konferentsiya molodykh uchonykh s mezhdunarodnym uchastiyem «Sovremennyye problemy i perspektivy razvitiya rybokhozyaystvennogo kompleksa». - 2019. – pp. 257-262. (In Russian).

[12] Lozhkin DM, Shevchenko GV. Cyclic Sea surface temperature variations in the sea of Okhotsk and adjacent water

areas according to satellite data in 1998-2018. Issledovaniye Zemli iz kosmosa. 2020;(1):44-51

[13] Rostov ID, Zhabin IA. Gidrologicheskie osobennosti priust'evoy oblasti reki Amur [hydrological conditions of the Amur River near-mouth area]. Meteorologiya i Gidrologiya. 1991;(7):94-99 (In Russian)

[14] Ivanov VV. Periodicheskie kolebaniya pogody i klimata [periodic weather and climate variations]. Uspekhi fiz. nauk. 2002;**122**:777-811 (In Russian)

[15] Ivanov VV. Issledovanie variatsiy srednemesyachnoy temperatury vozduha s pomosch'yu posledovatel'nykh spektrov [study of monthly mean air temperature variations with consecutive spectra]. Meteorologiya i Gidrologiya. 2006;(5):39-45 (In Russian)

[16] http://www.sakhniro.ru/

[17] Shevchenko G., Tshay Z., Puzankov C. EOF and wavelet analysis of satellite SST data in the northern Pacific // Proceedings of the Third PICES Workshop on the Okhotsk Sea and Adjacent Areas. Sidney, B.C., Canada, 2004. P. 13-18. (PICES Scientific report; N 26, 2004).

Chapter 11

Compression of High-Resolution Satellite Images Using Optical Image Processing

Anirban Patra, Arijit Saha, Debasish Chakraborty and Kallol Bhattacharya

Abstract

This chapter presents a novel method for compressing satellite imagery using phase grating to facilitate the optimization of storage space and bandwidth in satellite communication. In this research work, each Satellite image is first modulated with high grating frequency in a fixed orientation. Due to this modulation, three spots (spectrum) have been generated. From these three spots, by applying Inverse Fourier Transform in any one band, we can recover the image. Out of these three spots, one is center spectrum spot and other spots represent two sidebands. Care should be taken during the spot selection is to avoid aliasing effect. At the receiving end, to recover image we use only one spectrum. We have proved that size of the extracted image is less than the original image. In this way, compression of satellite image has been performed. To measure quality of the output images, PSNR value has been calculated and compared this value with previous techniques. As high-resolution satellite image contains a lot of information, therefore to get detail information from extracted image, compression ratio should be as minimum as possible.

Keywords: image compression, grating, Fourier transform, image retrieval, LISS- III sensor

1. Introduction

Compression of images is an important application in the field of satellite image processing as it is suitable for optimization of storage space and sharing over internet with optimum bandwidth utilization. For compression of satellite images, it is performed either directly from the image or from transformed part of the images. As discussed in [1] the compression of satellite images is based on Block Truncation Coding (BTC) technique. It first converts RGB satellite image into HSV planes. After that, each of the H and S planes are encoded using block truncation coding with quad clustering and V plane is encoded with BTC based bi-clustering or tri clustering depending on the edge information present in the plane. This method is better than previous BTC methods compared to visual quality of the output image. [2] discussed the image compression method based on evidence theory and k-Nearest Neighbor (KNN) algorithm. The main drawback is that the information loss is large [2]. To improve the quality of the output image, Fourier Transform and Huffman Coding is used for modification the previous technique. In both method, visual quality of satellite image is poor. [3] discusses the use of integer wavelet regression by increasing temporal correlation, which consequently improves the compression gain. [4] discusses a satellite image compression technique using discrete wavelet transform for noise removal to compress satellite images. [5] has discussed the use only hardware-based solutions in this lossless compression technique of X Sat images. [6] has analyzed the use of Discrete Wavelet Transform in their lossy image compression work and performance of different wavelets for satellite image compression. [7] has used the conventional Discrete Cosine Transform system for lossless image compression. [8] proposes an image compression technique using multiplexing and encryption by optical grating method [9–12].

In this chapter, we proposed a scheme to compress multiple high-resolution satellite images by using phase grating. Each image is modulated by applying high value of spatial frequency and a fixed orientation angles. For each image, multiple bands have been generated due to modulation which are placed in the same spectrum plane (only three bands are clearly visible). The spectrum is encoded and filtered using Gauss filtering. To detect the location of maximum image information, an intensity graph has been plotted in the decrypted plane. All stored images can be securely and efficiently retrieved by applying inverse Regional Fourier Transform operation. This proposed technique is simple and suitable for optimization of storage space and bandwidth in satellite communication.

The chapter is organized as follows:

- Section I describes the location and data used in this chapter
- Section II discusses the proposed compression method
- Section III presents proposed methodology of our research work
- Section IV provides result of our research work and PSNR value of the extracted images
- Section V concludes the paper and discuss why low compression ratio is desired for land cover analysis.

2. Description of location and data used

Images which are used in our research work, collected from Regional Remote Sensing Centre (East). The images are satellite picture of different areas in Kolkata Metropolitan Area.

All satellite images used in this paper are captured by LISS III 23 m sensor. LISS- III sensor is an optical sensor working in four spectral bands (Green, Red, Near Infrared and Short-Wave Infrared). It covers a 141 km- wide swath with a resolution of 23 meters in all spectral bands [13, 14].

3. Proposed methodology

3.1 Frequency and orientation angle selection for phase grating

According to rule of phase grating, value of the grating frequency ($u0$) should be high. Low grating frequency is creating aliasing problem and therefore it would be very difficult to reconstruct the original image. In this chapter, we propose to select

$u0 = 1400$, which is sufficient for filtering. In grating, value of the orientation angle (θ) varies from 0 to 360^0. We have worked with 0 deg orientation angle.

The diffraction gratings used are illustrated in **Figures 1** and **2**.

3.2 Mathematical expression

Let us assume that image, denoted by $f_1(x,y)$, is modulated by two sinusoidal phase gratings along the reference abscissa and of the form $\exp\left[i\left(\frac{m}{2}\right)\sin 2\pi u_1 x\right]$ where u_1 is the spatial frequency and m is the phase contrast. In such a case, the first modulated object may be expressed as:

$$s_1(x,y) = f_1(x,y)\exp\left[i\left(\frac{m}{2}\right)\sin 2\pi u_1 x\right] \tag{1}$$

Invoking the well-known convolution theorem, the Fourier transform of this modulated object is given by,

$$S_1(u,v) = [F_1(u,v) \circledast \left[\sum_{q=-\alpha}^{q=+\alpha} J_q\left(\frac{m}{2}\right)\delta(u - qu_1, v)\right] = \left[\sum_{q=-\alpha}^{q=+\alpha} J_q\left(\frac{m}{2}\right)F_1(u - qu_1, v)\right] \tag{2}$$

where J_q is the qth order Bessel Function of the 1st kind. The diffraction order is represented by the parameter 'q'.

The diffraction pattern, as given by Eq. (2) is a series of diffraction spots each containing the object spectrum. Considering the zero order and the first two orders of the spectrum, Eq. (2) may be represented by,

$$S_1(u,v) = \left[J_0\left(\frac{m}{2}\right)F_1(u,v) + J_{-1}\left(\frac{m}{2}\right)F_1(u + u_1, v) + J_{+1}\left(\frac{m}{2}\right)F_1(u - u_1, v)\right] \tag{3}$$

3.3 Retrieval of images from spectral band

The reconstruction of images is achieved by decrypting the encrypted plane followed by plotting an intensity graph of the spectrum horizontally. This intensity graph is generated using the intensity values with respect to its location (pixel position). In the intensity graph, peak values indicate the brightest spots, where information of images are maximum. To avoid any human intervention, system automatically finds the intensity level horizontally, vertically and diagonally. In addition, regional Inverse Fourier Transform is applied around the peaks by using a

Figure 1.
Sinusoidal phase grating along x axis.

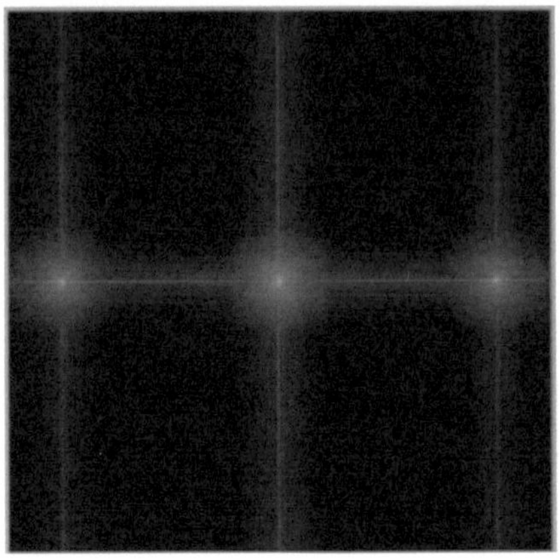

Figure 2.
Spectrum of an image due to modulation.

fixed cut-off frequency. As only one sideband is enough for image reconstruction, hence we select one side-band. This spectrum is filtering using a cut-off value of 120. Proper operating frequency is selected and there is no overlapping during the band selection, so the extracted images are free from aliasing problem. As center band contains the information of all images, therefore filtering is not applied across center band.

3.4 Quality checking of filtered image

PSNR value is calculated using Mean Square Error (MSE) approach:

$$PSNR = \frac{1}{mn}\sum_{y=1}^{m}\sum_{x=1}^{n}\left[f(x,y) - b(x,y)\right]^2 \quad (4)$$

Where $f(x, y)$ and $b(x, y)$ are original image and retrieved image respectively and m, n denotes size of images.

4. Results and discussions

Satellite Images chosen for testing the algorithm described in Section III are shown in **Figure 3(a)-(c)**. The dimensions of the selected images are 512 x 512.

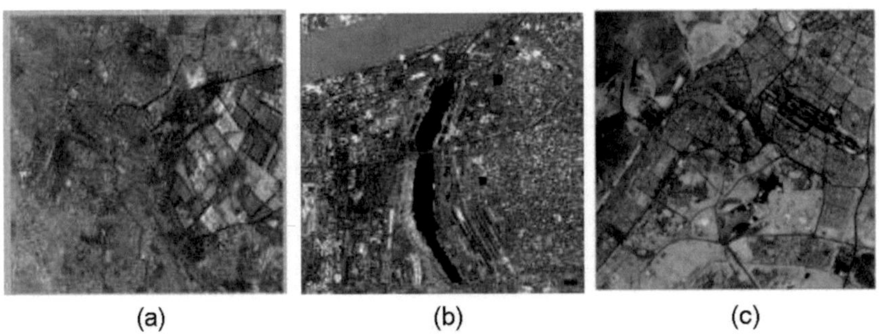

Figure 3.
Three high Resolution Satellite Images.

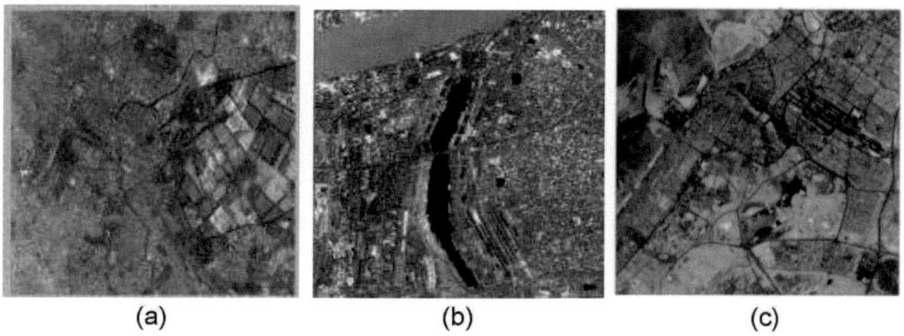

Figure 4.
Three extracted images.

Image	Original image size (kB)	Compression ratio (compressed image/original image)	PSNR
$f_1(x, y)$	242	0.84	32.5
$f_2(x, y)$	240	0.81	31.73
$f_3\ (x, y)$	254	0.86	31.6

Table 1.
Summary of PSNR calculation.

During zonal filtering operation, images $f_1(x, y)$, $f_2(x, y)$ and $f_3\ (x, y)$ have been extracted by Regional Inverse Fourier Transform taking upper spectrum from horizontal direction. Extracted images are shown in **Figure 4(a)-(c)**, respectively. A summary of the PSNR calculation is presented in **Table 1**.

5. Conclusion

In this chapter, phase grating technique has been proposed for compressing the high- resolution satellite images in frequency domain. The original image is retrieved by applying Inverse Fourier Transform from the respective spectrum of the image. As presented here, since we have taken only few coefficients from the spectrum, the size of the output image is less than the main original image. It should be mentioned that in high resolution satellite image, compression should be as minimum as possible. The main reason for the requirement of low compression ratio is mainly due to large geographical area representation (as these images contain a lot of information). High compression ratio is not suitable for accurate land cover analysis. To maintain the same dimension with original image and to avoid aliasing effect, spectral area is carefully selected. Compared with earlier methods, visual quality of the selected satellite images is very good as it is captured by optical LISS-III sensor. Our proposed technique is simple and suitable for optimization of storage space and bandwidth in satellite communication.

Author details

Anirban Patra[1*], Arijit Saha[2], Debasish Chakraborty[3] and Kallol Bhattacharya[4]

1 Department of ECE, JIS College of Engineering, Kalyani, West Bengal, India

2 Department of ECE, BPPIMT, Kolkata, West Bengal, India

3 RRSC (East – A Unit of ISRO), Kolkata, West Bengal, India

4 Department of Applied Optics and Photonics, University of Calcutta, West Bengal, India

*Address all correspondence to: anitublu@gmail.com

References

[1] B. Balakrishnan, S. H. Darsana1, J Mathews, M S. Nair; "Satellite/Aerial Image Compression Using Adaptive Block Truncation Coding Technique"; Journal of Indian Society of Remote Sensing, July 2018

[2] K Sahnoun, N Benabadji ; "Satellite Image Compression Technique Based On The Evidence Theory"; Advanced Computing: An International Journal (ACIJ), Vol.5, No.1, January 2014

[3] K Sahnoun, N Benabadji ; "Satellite Image Compression Algorithm Based On The FFT" ; The International Journal of Multimedia & Its Applications (IJMA) Vol.6, No.1, February 2014

[4] Md. Al Mamun and Md. Ali Hossain ; "Satellite Image Compression Using Integer Wavelet Regression" ; International Conference on Electrical, Computer and Communication Engineering (ECCE) ; 2017

[5] K Sahnoun, N Benabadji ; "Satellite Image Compression Technique Using Noise Bit Removal and Discrete Wavelet Transform"; International Journal of Imaging and Robotics; Volume 15, Issue Number 3 , 2015

[6] R.M. Susilo , T.R. Bretschneider; "On the real time satellite image compression of X-Sat" ; Fourth International Conference on Information, Communications and Signal Processing, 2003

[7] T Memane and S D Ruikar ; "Selection Of Wavelet For Satellite Image Compression Using Picture Quality Measures"; International Conference on Communication and Signal Processing ; 2014

[8] I. Hacihaliloglu and M. Kartal ; "DCT and Wavelet Based Image Compression In Satellite Images"; Recent Advances in Space Technologies, 2003. RAST'03.

[9] A Patra, A Saha, K Bhattacharya ; "Multiplexing and encryption of images using phase grating and random phase mask" ; Optical Engineering 59 (3), 033105-033111; 2020

[10] A Patra, D Chakraborty, A Saha, K Bhattacharya ; " Compression of Satellite Images using Sinusoidal Amplitude Grating": ; IJEE 11(1); 2019

[11] A Patra, S Bandyopadhyay, D Chakraborty, A Saha ; " A Novel Approach to Compression of Satellite Images Using Butterworth Filtering"; Springer LNSS; 2020

[12] A Patra, D Sutradhar, S Nath, S poddar, R Nandi ; " Study on Video Compression and Reconstruction using Downsampling and Upsampling"; International Journal of Advance Research in Science and Engineering 9(5); pp-12-17; 2020

[13] Madhubala M , S.K.Mohan Rao, G. Ravindra Babu; Classification of IRS LISS-III Images by using Artificial Neural Networks; IJCA Special Issue on "Recent Trends in Image Processing and Pattern Recognition" RTIPPR, 2010.

[14] R Naidu, M.V.S.S Giridhar ; Un-Supervised Classification of Rice Crop using IRS LISS III Satellite Images for Wazirabad Command Area; International Journal Of Engineering Development And Research